Cambridge Elements ≡

Elements in the Philosophy of Biology
edited by
Grant Ramsey
KU Leuven, Belgium
Michael Ruse
Florida State University

BIOLOGICAL INDIVIDUALITY

Alison K. McConwell
University of Massachusetts Lowell

CAMBRIDGE
UNIVERSITY PRESS

Shaftesbury Road, Cambridge CB2 8EA, United Kingdom

One Liberty Plaza, 20th Floor, New York, NY 10006, USA

477 Williamstown Road, Port Melbourne, VIC 3207, Australia

314–321, 3rd Floor, Plot 3, Splendor Forum, Jasola District Centre,
New Delhi – 110025, India

103 Penang Road, #05–06/07, Visioncrest Commercial, Singapore 238467

Cambridge University Press is part of Cambridge University Press & Assessment,
a department of the University of Cambridge.

We share the University's mission to contribute to society through the pursuit of
education, learning and research at the highest international levels of excellence.

www.cambridge.org
Information on this title: www.cambridge.org/9781009387422

DOI: 10.1017/9781108942775

First published 2023

A catalogue record for this publication is available from the British Library.

ISBN 978-1-009-38742-2 Hardback
ISBN 978-1-108-93189-2 Paperback
ISSN 2515-1126 (online)
ISSN 2515-1118 (print)

Cambridge University Press & Assessment has no responsibility for the persistence
or accuracy of URLs for external or third-party internet websites referred to in this
publication and does not guarantee that any content on such websites is, or will
remain, accurate or appropriate.

Biological Individuality

Elements in the Philosophy of Biology

DOI: 10.1017/9781108942775
First published online: May 2023

Alison K. McConwell
University of Massachusetts Lowell

Author for correspondence: Alison K. McConwell, alison_mcconwell@uml.edu

Abstract: This Element develops a view about biological individuality's value in two ways: while biological individuality matters for its theoretical and methodological roles in the production of scientific knowledge, its historical use in promoting the politics of social ideologies concerning progress and perfection of humanity's evolutionary future must not be ignored. Recent trends in biological individuality are analyzed and set against the history of evolutionary thought drawing from the early twentieth century. This title is also available as Open Access on Cambridge Core.

Keywords: Individuality, history, practice, progress, eugenics

ISBNs: 9781009387422 (HB), 9781108931892 (PB), 9781108942775 (OC)
ISSNs: 2515-1126 (online), 2515-1118 (print)

Contents

The Purpose of This Element

This Element analyzes interdisciplinary and philosophical discussions of biological individuality. For philosophers, biological individuality is a problem space both old and new. The problem of individuality occurs across numerous disciplines and is wrapped with notions of identity, time, categories, nature, and quite frankly ourselves and what makes us who we are. The life sciences continue to provide exciting puzzles challenging intuitions about how nature is organized, and in turn, how we use concepts to organize nature.

Yet, some have challenged whether biological individuality matters in the production of scientific knowledge and its usefulness as a topic more generally: Why does individuality matter for biology? For philosophy? In other words, *what is its value?*

There are two ways philosophers tend to think about values in science. One concerns epistemic, or "knowledge-based" values about reasoning, method, theory, success, and characteristics of how knowledge is attained. And so, Sections 1 and 2 of this Element focus on the theoretical and methodological aspects of biological individuality, and its role in the production of scientific knowledge. The second way philosophers consider value concerns social and political features, often called "non-epistemic values." Section 3 takes that non-epistemic (i.e., social and political) turn.[1] The non-epistemic value of biological individuality has been under-explored. By drawing from naturalists like Darwin, the Huxleys, and Asa Gray in the history of evolutionary thought, I argue that biological individuality promoted politics of social ideologies about managing the direction of human evolution with the life sciences.

In that sense, I submit that biological individuality is not, and never has been, value-free. Biological individuality's dark side serves as a cautionary tale; the concept is shaped by social and political ideologies about progress and perfection.

The following contains a series of essays meant to inform those new to the problem of biological individuality. The aim is to analyze recent trends against select histories of evolutionary thought, specifically around the early twentieth century.

To the experts, many of whom are cited in these pages, a single Element on this topic cannot apply across all contexts nor comprehensively capture the details of every intellectual endeavor worthy of analysis. This Element is designed for accessibility to students and junior scholars, but it also aims to contribute to the intellectual arena. The sections are structured accordingly.

---

[1] The distinction between epistemic/non-epistemic values is a useful heuristic but rationality and reasoning are not devoid of social features; non-epistemic (or perhaps better "contextual") values matter for knowledge in Longino's sense.

Section 1: An Ontic Landscape maps the ways biological individuality is theoretically and conceptually defined according to the life sciences. "Ontic" refers to what exists (i.e., objects, concepts, categories, properties, etc.) in a domain (i.e., physical, chemical, biological, but also subspecialty domains like immunological, ecological, etc.). I call these approaches "domain-driven" because their analyses derive from select disciplinary domains or subspecialties including evolutionary biology, immunology, ecology, and so on. Certain domains have received more attention than others. As we'll see, domain-driven approaches yield many (sometimes non-evolutionary) ways to define biological individuality, and that resulting plurality and its ambiguities must be sorted and discussed.

Section 2: Critics & Methodology. Critics of work discussed in Section 1 ask epistemic questions like what value, if any, biological individuality has in producing empirical knowledge. These critical approaches I take to be "practice-based"; attention directs to how biologists, working in lab and field contexts, use and think about biological individuality. And so, I distinguish three types of practice-based approaches, which include how individuality concepts function in producing empirical results. Further, preoccupations with phenomenal qualities of biological objects – for example, what those objects are like in terms of how their boundaries are distinguished from their environments – is critically analyzed. Recommendations are provided for newcomers to avoid a cottage industry of this topic. Philosophers must avoid remanufacturing standard puzzle cases against received concepts of biological individuality. In light of that critique, Section 2 closes with a new opportunity for philosophical analysis at the cross-section of philosophy, biotechnology, and values.

Section 3: In Historical Context. Biological individuality has a long (and fraught) history outside of analytic philosophy, a history led by naturalists of the nineteenth and twentieth centuries. The historical figures in this section are anything but obscure in the history of biology: they wrestled with notions of agency, design, perfection, and progress in their disputes with the church concerning intellectual authority over nature. While Sections 1 and 2 focus on theoretical and methodological aspects of individuality's value for gaining knowledge about the biological world, Section 3 takes a social and political turn showcasing biological individuality's social significance. I argue that biological individuality was used to promote political and social ideologies about managing the "perfection" of human evolution. There are not only theological features, but alarming eugenics-overtones harnessing biological individuality as a tool for control over humanity's evolutionary future.

I hope *Biological Individuality* will reveal new ways for readers to think about individuality, while also revisiting places some readers know well.

As a graduate student, I found the topic very complicated and difficult. The sections of this Element are written in a way that draws from what I wish I would have known and where I hope to see work go in the future. Biological individuality is anything but a trivial conceptual space both in the concept's complexity and its relevance for philosophical and scientific debates.

I invite all readers to make this Element their own. While shaped by an overarching thread of argument concerning biological individuality's value, sections can be approached by prioritizing different routes of investigation.[2] However, all three sections are intended to cohere such that each carries a sense of belonging and function taken all together as one single individual Element.

1. An Ontic Landscape

Introduction to Section 1

> Life in general consists of the life-histories of individuals.
> —Child (1915, 5)

Upward of 30 trillion human cells are outnumbered by approximately 39 trillion bacterial cells. Some cells, for example, microbes in the gut and brain, are capable of altering behavior and neurotransmitter levels (Sampson and Mazmanian 2015). In what sense, then, are humans *individuals* in their own right, rather than merely part of a greater microbial complex? Some argue that a symbiotic view of life, one prioritizing interactional relationships among and between organisms and their microbes, reveals that humans have never been individuals (e.g., Gilbert et al. 2012). What exactly are biological individuals and why do they matter for the biological sciences? And how might philosophers develop answers to such questions?

Challenges like the above case invite exploration of traditional philosophical terrain informed by empirical disciplines. Disciplines are distinguished by their domain of subject matter. Broader domains, like the life sciences, can include subspecialties meaning that individuality concepts, like evolutionary,

---

[2] Nelson Goodman's 1978 *Languages of Art* inspires the structure of this book with different possible routes of investigation. Readers may prioritize historical analysis in Section 3 before ontological and methodological analyses in Sections 1 and 2. Alternatively, readers may start with Section 2's methodological focus before reading the theoretical and historical works in Sections 1 and 3. In contrast, standard linear reading develops a narrative about biological individuality's value through theory (Section 1), practice (Section 2), and history (Section 3), which for conventional reasons prioritizes a theoretical survey and analysis, followed by methodological critique, and finally a historical analysis to contextualize biological individuality's social and political value.

immunological, ecological, and metabolic individualities, are each defined and understood according to their own domain of study. That is what it means to develop an ontic landscape as *domain-driven*: conceptual analysis is theoretically and conceptually derived from disciplinary specialties in the life sciences. For example, evolutionary individuals are discussed in relation to evolutionary biology, which are contrasted against individuals relevant to other areas, such as immunology.

This section surveys recent disputes developing a pluralistic approach to biological individuality. Organismality as an organizing principle is discussed first, then species as individuals. How evolutionary individuality expanded reproduction's conceptual scope is also considered. After, individualities in non-evolutionary contexts, such as immunology and ecology, are analyzed to demonstrate biological individuality's theoretical value to matters of life and health.

There are many types of biological individuals. While biological individualities are categorized and classified according to a domain, there are different approaches to pluralism that must be sorted. I develop that pluralism both synchronically and diachronically; there are many types of biological individualities both at a time and over time. As a reference tool, the appendix (Table A.1) gathers several cases discussed throughout this Element from clonal organisms, to eusocial colonies, to social amoeba and more.

Let's start with organisms.

Organisms

'Individual' and 'organism' were once synonymous terms (see Buss 1987). However, organismality is now considered one organizational category under the umbrella of Individuality. In what follows, organismality is explored according to historical considerations, conceptual contrasts, and etymological analysis. After, key takeaways are provided about organismality's epistemic value as an organizing category.

1. Historical Considerations

Organisms were, at one time, the best representatives of individuality. First, consider how naturalist and evolutionist Julian Huxley professes his views to the philosopher in the preface of his book *The Individual in the Animal Kingdom* (1912):

> Living matter always tends to group itself into these "closed, independent systems with harmonious parts." Though the closure is never complete, the independence never absolute, the harmony never perfect, yet systems and tendency alike have real existence.

Huxley believed that organisms were more individuated than nonanimate crystals. Organism boundaries were definite: their size and form were defined by a scheme of architecture in contrast to inorganic crystal systems growing without limits. Organisms were more independent in their self-determining qualities (1912, 51). That is, their agency – their capacity to self-sustain and repair against perturbations – is what made organisms proper objects of biology. For Huxley, they were not only the best representatives of individuality, but organisms were central to navigating differences among organic and inorganic materials.

In contrast, consider the physician Sir William Osler's Ingersoll Lecture (1904) when he discussed the meaning of death against lessons of embryology:

> The individual is nothing more than the transient off-shoot of a germ plasm, which as an unbroken continuity from generation to generation, from age to age . . . "the individual organism is transient, but its embryonic substance, which produces the mortal tissues, preserves itself imperishable, everlasting, and constant".

Osler is not denying organismality's existence, but rather the *significance* of individuatedness it's supposed to represent. Osler's view draws from Weismann's germ-soma distinction identifying the germ plasm as central to heredity across generations. He isolated the germline from developmental events of the individual organism's life cycle (Richmond 2001, 169). Organismality for Osler, then, is ontologically secondary to the eternal generational thread.

The historical considerations above yield one lesson about organismality's significance: Huxley prioritized organismality as an entry point to access relevant features of individuality for *life's* evolution (versus changes in nonliving, inorganic material). However, Osler emphasized continuity of genetic lineages for which organismality was just a vessel. As products of their time in the early twentieth century, for Huxley and Osler organismality was representative of individuality, yet as an organizing principle it functioned differently in their approaches.

2. Conceptual Contrasts

How organismal parts work together in the larger system sustaining life was historically conceptualized in relation to structural constitutions of inorganic systems, like crystals and "habits" of minerals exhibiting change and structural order. What distinguished organisms as alive prior to nineteenth century biology was the unobservable, nonmaterial substance *elan vital* or "the force of life." However, a post-Newtonian scientific world demanded rejection of mysterious qualities in favor of mechanical descriptions, that is, in favor of how parts function together to produce system-level effects.

Organisms as complex systems were analyzed into component parts by different naturalists including anatomists, physiologists, embryologists, and so on (see Hull 1978, 336). Ruse (1987, 225) argued that individual organisms can be fragmented into structurally various parts functioning together interdependently to sustain the entire unit. At the same time, he acknowledged the complexities of decomposing organisms into discrete characters based on function and ancestry. In an evolutionary context, decomposition matters for building phylogenies and classifying taxa, which sort organisms across the Linnean hierarchy.

However, decompositional approaches are often contrasted with holist goals. Nuño de la Rosa (2010, 290) explains that Organicism – a holistic tradition regarding organisms defined as functionally-integrated and autonomous systems – has more ancient and historical roots than Darwinian theory. But at least two traditions can be distinguished for conceptually analyzing organismality.

On the one hand, under Darwinian traditions in the shadow of modern synthesis orthodoxy, "organisms are included in the more general category of biological individuals, defined as those entities (not only organisms but also genes or species) on which natural selection acts" (Nuño de la Rosa 2010, 290). Continuing the critique of organisms as mere vessels of adaptative characters: "organisms are conceived of as a non-problematic kind of individuals composing populations, and their distinct parts [their characters] are abstracted as adaptive traits that assure [an organism's] reproductive success within specific environments" (290). In other words, organisms matter for more than their role as adaptation bearers, a role that atomizes and isolates parts as theoretically primary.

On the other hand, Nuño de la Rosa argues that in fact there are non-evolutionary morphological or physiological theories that prioritize organisms as integrated wholes through their developmental lifetime. By appeal to organicism's longstanding history of varied views emphasizing connectedness and integration, she argues that strong theoretical grounds persist from Aristotle and Kant to the experimental embryology and developmental biology of the late nineteenth and twentieth century.

Sometimes the organism concept is used to synthesize intellectual traditions just discussed. For example, Huneman (2017) offers a conception of organismality to support evo-devo traditions combining developmental and adaption-focused views. One maps onto epigenetic self-production of parts within a viable whole, and the other explains design of the whole by natural selection. In sum, organismality has been conceptually considered according to approaches that decompose organisms into their adaptive character traits, approaches that consider their developmental features as living cycles, and combined approaches.

3. Etymological Analysis

Finally, "[o]rganisms are so called because they are literally *organized*" (Simpson 1958, 519). The term 'organism' has a long history. Etymology reveals 'organic' in reference to natural organization occurred around the late 1600s to early 1700s. The suffix 'ism' denotes a distinctive practice or system of some kind: organ-*ism* in its literal sense refers to a form of organization adapted for use in natural (i.e., non-artificial) contexts. Cheung (2006, 319) traces first appearances of the term in the life sciences and its usage in different settings. In the later 1700s, 'organism' became an ordering principle and a "generic name for individuals as natural entities or living beings" (2006, 319). However, *living* order as a mechanical product of an organism's parts working together needs more historical context.

Historian Jessica Riskin explains that the ancient model of living machinery persisted through the medieval Scholastics. By the mid-1600s it was as familiar as "automata on clocks and organs in churches and cathedrals" (2018, 159). When Descartes wrote the *Treatise of Man* in the 1630s, an anatomical treatise, he applied a different method from his predecessors in ancient and medieval anatomy (2018, 144). Riskin states that the analogy about mechanistic clockwork,

> . . . did not imply that the phenomenon in question [organismality] resembled a clock. It meant rather that the comprehension to be achieved was comparable to a clockmaker's understanding of a clock .... . clockwork meant intelligibility in terms of material parts, not literal clockwork. Descartes's animal-machinery resembled ancient and medieval animal machinery in many respects: it was warm, fluid, responsive, mobile, sentient, and full of agency. Its salient difference was that it was fully material and so completely intelligible in Descartes' new science (2018, 147).

And so, the intelligibility of organism function was realized in terms of its material parts – its anatomy, which was *not* Cartesian machinery in the pejorative sense.

Three vantage points, historical, conceptual, and etymological, were just considered as depicted in Figure 1 below. So, organismality's epistemic value as an organizing principle can be summarized as follows.

First, organismality was a conceptual lens for understanding how living (versus nonliving) systems function. Organismality was considered in contrast to inorganic systems like crystals and compared with artificial systems and machines.

Organismality: A Brief Garden Walk Through An Organizing Concept

Historical Considerations	Conceptual Contrasts	Etymological Analysis
A primary entry point for understanding life's changes (vs. change in non-living, inorganic change)	Decomposing mechanics of parts, isolating characters as evolutionary adaptations as theoretically primary	• Late 1600s/early 1700s: 'Organic'—> natural organization • Adding 'ism': Distinctive form of organization • End of 1700s: 'organism' denoting natural, living individuals (Cheung 2006)
Vs.	+	+
A secondary vessel for understanding continuity through change (mere bearers of life's "eternal generational" thread)	Emphasizing integrated, system level morphology, physiology, developmental features as theoretically primary	Additional Historical Context: Organization intelligible in terms of material parts (clockmaker's understanding of a clock translated to scientist's understanding of an organism)

Key Take Aways

1. Conceptual role in building knowledge of living (vs. non-living) systems
2. Shaped biology's early scientific status in post-Newtonian framework
3. Serves as different theoretical entry points: primary/secondary

Figure 1 Summary of organismality as an organizing concept

Second, organismality's epistemic value is evidenced by its role in shaping biology's early scientific status in a post-Newtonian era. Mechanical function and decomposition of machines informed and constrained analysis, even organicist critiques that challenged the mechanical–vitalism dichotomy. Mysterious qualities to explain organisms as *living* systems were rejected in both decompositional and organicist accounts.

Third, organismality served as both a primary and secondary analytic entry point: for prioritizing quasi-closed and autonomous systems as agents of change (as per Huxley) and as a mere vessel for continuity (as per Osler's rendition of the eternal thread). However, there's more to biological individuality than organisms alone.

Individuality, Classes, & Species

Species taxa consist of organisms grouped together in a particular way. There are numerous species concepts to group organisms into species (e.g., Mayr's 1970 interbreeding and Van Valen's 1976 ecological approaches). In the latter twentieth century, debate ensued over the metaphysical nature of species: What *is* a species? Are species like classes akin to chemical kinds on the periodic table? Or something else?

The species-as-individuals thesis or S-A-I is the view that species taxa are not classes or kinds, but instead individuals. The following centers on David Hull's 1965, 1976, 1978, and 1980 papers. One thread of Hull's work concerns an argument by analogy: he identified features of organisms representing their individuality, which he then extended to the case of species because species share those same features. That is, if organisms are individuals because they are cohesive, discrete, spatiotemporally restricted entities with beginnings and endings in time, and species have those features, then species are individuals too. According to Hull, organisms and species are similar: they satisfy criteria of *metaphysical* individuality.[3] However, why argue that species are individuals? Motivations of S-A-I are traced before assessing (and rejecting) some interpretations of Hull's work.

1. Tracing the Motivations of S-A-I

Why argue that species are individuals? The S-A-I thesis was proposed in response to ancient, pre-Darwinian views that species are static categories of nature.

As Ereshefsky (2022) points out, since Aristotle species have been the main examples of natural kinds (i.e., of natural categories independent of our classification schema) with essences. In pre-Darwinian contexts, species were created (by the gods or later, God), each endowed with essential characteristics – an essence – signaling species membership. Classifying species taxa occurred by shared qualitative characteristics, which were unique to a species and necessary in that all and only members of a species have them. However, even Linnaeus had difficulty determining a species' essence, and evolutionary theory explains why: Forces like selection, mutation, recombination, and random drift can cause traits to disappear over time (Ereshefsky 2022, 2001).

Early on, Hull (1965) explained negative effects of essentialism on taxonomy, what he called "two thousand years of stasis" in response to Ernst Mayr's paradox. Mayr pointed out that while taxonomists accepted evolution, they still adhered to conceptualizing species as static entities. Hull (1965, 316) worked to unpack what he and others viewed as the problem's crux. Essentialism's residue was responsible for the conflict taxonomists faced. In particular, Mayr's paradox was due to essentialist views of species as natural kinds or classes defined by shared essences precisely because evolutionary change precludes species taxa as static, unchanging entities.

Hull (1978) contrasted metaphysical notions of natural kinds and classes with individuals. Classes are groups of entities that can function in scientific laws, whereas individuals are historical entities that occupy particular space-time regions (1978, 337). Members of a certain class belong to that class because of the attributes they share. In modern contexts, most common examples to

---

[3] An expansion of concepts marked the 1970s: the organism concept was controversially used by James Lovelock and Lynn Margulis to describe the earth itself as a single living "organism" known as the Gaia hypothesis. So, the nature of organismality was conceptually expanded across levels of organization. But if individuality is supposed to be distinguished from organismality by *individuality's* expansion across levels of organization, then their relevant differences remain under-explored.

illustrate natural kinds and their essences draw from chemical kinds on the periodic table, for example, all instances of gold have the atomic number seventy-nine. Consider the following three features of classes.

First, classes serve nomothetic aims by providing a stable, reliable base for induction; laws generalize over features of classes; reliable inferences can be made about how members behave under certain conditions. For example, pure gold melts at 1948 degrees Fahrenheit in standard atmospheric pressure, pressure which is defined at sea level. The melting point of gold is reliably inferred not only by its chemical constitution but also by how that constitution behaves under specific conditions.

Second, classes are spatiotemporally unrestricted or "forever open" meaning that members can in principle re-appear at different times and places, whereas individuals are spatially and temporally located with beginnings and endings in time.

Third, members of classes share similar attributes and do not exist in part–whole relationships with other members of their class. Parts of an individual need not be similar, for example, an individual organism can be fragmented into structurally various parts that function together interdependently to sustain an entire organism (Ruse 1987, 225).

So, if species are *not* classes, this implies: (1) It's possible for inferences to fail. There is no guarantee for species behavior (i.e., genetically, morphologically, or behaviorally) in certain conditions. (2) The same species cannot go extinct and re-emerge later because species taxa are unique to specific times and places.[4] And (3) not all organisms in a species will necessarily share an essential "core" set of attributes.

Later, Ruse (1987) argued that most philosophers discussing species as natural kinds were not in touch with biological reality. While Ruse raised objections against the S-A-I thesis, he clarifies motivations behind S-A-I. Typological views infused with static isolation and unchangeability dominated pre-Darwinian thought about species. We want to say that species are real, Ruse argues, but also that they can change. And so, one theoretical motivation driving S-A-I was its promise to designate species as tangible, concrete, *and* changing entities.

While Hull's 1965 paper characterized the problem of species as natural kinds in light of evolution, that was one year after S-A-I's initial formulation emerged from Ghiselin (1966, 208–209) who proposed that biological species are "in the logical sense" individuals. He argued that to think otherwise is

---

[4] For Hull (1976, 184), the individuality status of species meant that "the same species can no more re-evolve than the same organism can be born again."

a category mistake. Species names are proper nouns and species must be individuals, metaphysically speaking, in order to evolve. Individuals, Ghiselin (1987, 128) claimed, are single things, including compound objects made up of parts, while classes, as previously defined, are unrestricted to definite locations in space and time, so their names may designate any number of objects – including none at all. Hull (1976) took Ghiselin's view further.

2. Hull on S-A-I: Interpretation & Clarification

With motivations in hand, it's time to analyze Hull's 1976, 1978, and 1980 papers. After some interpretations are proposed and criticized.

One important distinction from "Are species individuals?" (1976) concerns the species category versus species concepts. Species *concepts* organize species taxa according to some set of criteria, while the species *category* is defined in contrast to other Linnean classifications like genus, family, and order. Consider Figure 2 below.

According to Hull (1976, 174) the old view of species defined the species category as a class *of classes*, that is, species taxa were classes or natural kinds. He did not contest the species *category* as a class but instead outlined how species *taxa* are individuals in the same way organisms are. He says, "the relation which an organ has to an organism is the same as the relation which an organism has to its species" (1976, 181). Evolution as a selection process requires relations among organisms in a species to be one of *continuity*: organisms reproduce themselves over time and as such are integrated into species, as historical entities, by descent. This why species taxa cannot be classes so long as classes remain ahistorical; class membership is unrestricted by history of time and place. For example, for something to be gold, it does not depend on its location along the American River in Coloma, California in the mid nineteenth

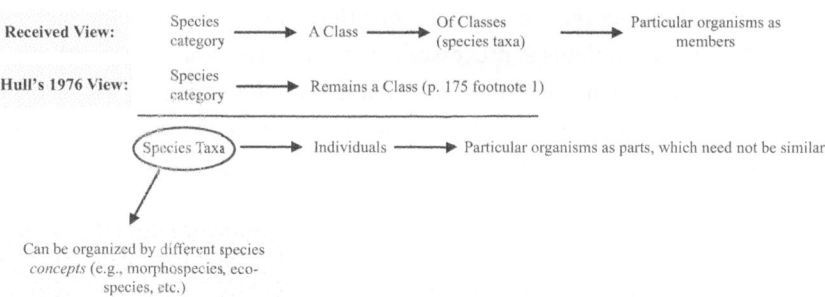

Figure 2 The received view that species are a class of classes versus Hull's development of Ghiselin's view that species taxa are individuals

century – gold membership is unrestricted by time and place, and thus discontinuous. Consider the following reconstruction of Hull's 1978 argument:

1. Evolution requires historical entities
2. Classes are not historical (they are spatiotemporally unrestricted)

   _____

3. Species cannot be classes
4. Individuals are historical entities by definition

   _____

5. Species are individuals

Species taxa, the basic units of classification, are the basic units of evolution – *individuals* in evolution – because they are historical, which made the search for qualitative similarity in essentialist approaches to classification a red herring (1978, 348). Species taxa are individuated from one another on the basis of continuous descent and cohesiveness, and not by similarity or essence (355–356).

While historical continuity is a necessary condition for individuality, maintaining a sufficient level of unity or cohesiveness is also required to form an individual unit. How Hull thought of unity at this juncture is complicated: while some entities have sufficient unity to compete with one another but not the open-ended organization to evolve, such as organisms, others are capable of open-ended change but might lack sufficient unity, such as higher taxa (184). While sufficient unity seems underspecified, intrinsic and external causes are identified as responsible factors (183– 184). For example, if asexual forms specialize, become adapted, and shift their adaptations, then asexual species lack intrinsic mechanisms for promoting their evolutionary unity, and as such form units entirely by the unifying effects of external causes (183).

Regardless, individual units are the objects of processes, and so one might presume that genes are *the* unit of mutation, organisms *the* unit of natural selection, and species *the* unit of evolution (181). However, Hull acknowledges complex views of biologists: processes can occur across wide ranges of organizations. Mutations can consist of single nucleotide alternations or as the loss or gain of entire chromosomes. Plus, Lewontin (1970) argued that natural selection occurs across levels from macromolecules to populations. So, Hull clarifies in the later 1980 paper:

> Entities at various levels of organization can function as units of selection if they possess the sort of organization most clearly exhibited by organisms; and as such, units of selection are individuals ... like mutation and selection, evolution occurs at more than one level of organization (1980, 131).

Hull argued that entities at various levels of organization (e.g., genes, cells, organisms, species, etc.) can function as units of selection, which implies that natural selection occurs at and across various levels.[5]

Species are individuals for Hull, though, because they are units of *evolution*: to be units of evolution species must form lineages where natural selection causes those species to evolve (see Ereshefsky 2022). In other words, species form lineages of "evolutionary unity" that can be individuated by spatiotemporal location and continuity (1978, 344). As historical entities, species form lineages just like genes and organisms form lineages, and they persist "while changing indefinitely through time" (341). Hull briefly distinguishes between units of selection and units of evolution, but he thought that *both* are individuals (1978, 338). In other words, species as units of selection was a contentious idea, but even if one denies species are units of selection, they are still spatially and temporally continuous, and as such are historical entities that evolve as a result of selection at lower levels.[6]

So, the term 'evolutionary individual' can be ambiguous if we do not necessarily mean the objects of natural selection, but rather units of evolution. As we'll see, contemporary usage of 'evolutionary individual' in the next section concerns individuals in natural selection. However, Hull's evolutionary individuals (e.g., with species as the exemplar evolutionary units) are entities that become adapted or shift their adaptions, and not the bearers of adaptation like individuals in selection are. Whether species can be objects of selective processes – as both individuals of evolution and *of selection* – was the point of contention though.

Rejecting Alternative Interpretations. One reviewer noted that Hull's 1976 paper is explicitly an ontology of the species category, and that it would be misguided to emphasize Hull's earlier work as not focusing on the species category's ontological status. To this point, I disagree. Hull did not explicitly focus on the species category, rather his focus is that species *taxa* are, ontologically speaking, not classes but individuals. Hull concedes (to Ernst Mayr in footnote 1 on p. 175) that the species category remains a class, however, he doesn't provide a positive argument in that paper for *why* the species category

---

[5] Hull (1978, 338) identifies need for conceptual clarification about evolutionary individuality apart from the levels of selection debate. It's not about the level on which selection operates: the nature of evolutionary individuality is identified by characteristics satisfied across multiple levels of organization.

[6] Hull's conversation with Simpson across the 1976 and 1978 papers has been under-analyzed. Hull identifies the old, gradualist view wherein species change indefinitely through time as paradigm lineages resulting from micro-level processes. However, if species do not change much in the course of their existence (per Gould and Eldredge), they cannot evolve, but instead form lineages (e.g., like genes and organisms do) and those lineages evolve (1980, 327).

remains a class. In his own words, Hull strategically accepts that his analysis does not affect the species category's class status:

> Ernst Mayr pointed out the need to emphasize the fact that the species category remains a class on the analysis being presented in this paper (175).

This means that the species category and species taxa are treated differently: Hull remarks that as a class, the species category can continue to be identified in the "usual," that is, *non-contested*, way (175).[7] The usual, received view of the species category was that it is a class. This leaves open the possibility that, if life exists on other planets, the species category would apply there as well. In contrast, and given the historical nature of evolution, one could explore the species *category* as itself an individual too pertaining to the tree of life on Earth specifically (and not on other planets). However, opening that pandora's box is a task for another day.

Additionally, that same reviewer claimed that Hull described three kinds of evolutionary individuals (in the 1976 and 1978 papers) – units of mutation, selection, and evolution – by using the method of theoretical individuation. By identifying the role these types of evolutionary individuals play in theory, namely, three theoretical roles, there are three kinds of evolutionary individuals. Furthermore, Hull (1980) then distinguishes units of selection into two types: replicators (i.e., entities that retain their structure largely intact through descent) and interactors (i.e., entities that cohesively interact with their environment in a way that has a unitary effect on constituent replicators).[8] So, they summarize, there are four types of evolutionary individuals for Hull: units of mutation, two units of selection, and units of evolution.

First, I find this characterization of how Hull views ontology and theory to be too simplistic, or at least it must be drawn out. It's well known that in 1992, Hull grounded his evolutionary treatment of biological individuality on a theoretical basis (i.e., the theory of evolution by selection) in the absence of competing physiological or morphological theory.[9] However, Hull is explicit in 1980 that he focuses on characteristics of processes, and how entities perform with regard to those processes. He identifies "ontological status" as referring to the differences between class-inclusion, class-membership, and part–whole relations (1976, 181, footnote 6). Specifically, ontological status concerns the relations between entities as they pertain to those logical types. He says, "ontological

---

[7] Concerning species concepts and the species category, unification and realism do not necessarily go hand in hand. The species category is a heterogenous category of very different types of entities (Ereshefsky 2000, 147, 157)

[8] Hull (1980, 316) distinguishes his replicator/interactor model from that of Dawkins' replicators that were characterized as passing on *identical* structure with passive vehicles as their containers.

[9] Nuño de la Rosa (2010) thinks Hull was too quick to draw that conclusion. We return to that later.

status is theory-dependent" (in that same footnote) because he's referring to evolution as requiring species taxa to be historical entities, which classes cannot be. What the theory of evolution determines is that species taxa cannot be classes, which leaves the part–whole relations of individuals as the better logical type.

Units of mutation, selection, and evolution across levels and as objects of processes are not merely manifestations of theoretical roles. Ontology – by way of logical relation types listed above – is theory-dependent for Hull because, as he says, nature does not come with logical types written on its face. Evolutionary processes are not theoretical posits, and so the objects of those processes are not either. Even when distinguishing two units of selection (i.e., replicators/interactors), Hull (1980) is consistent in that he *starts with* processes. In that regard, function is not merely a theoretical role that defines categories from normative expectations set by theory.

And second, what I take to be less important is the number of types (e.g., 3 or 4) to be distinguished from Hull's work, at least for the purposes of what motivates S-A-I and why it matters. One significant motivation for Hull is that evolutionary theory demands species taxa be historical entities, rather than ahistorical natural kinds viewed as unhelpful pre-Darwinian residue. Hull was also engaged in consistent conversation with paleontologists like Simpson, Gould, and Eldredge over the status of species change in particular: whether species change indefinitely through time as merely the result of lower-level selection or whether species taxa form lineages, lineages which themselves evolve through macrolevel processes (e.g., see 1980, 327). Hull did not endorse static types in the case of species taxa, and it is doubtful he would endorse its analog in the case of individuality.

In closing, S-A-I has normative impact later articulated by Hull concerning human species membership. Throughout history, many people were de-humanized as deviants from humanity. In response, Hull's view implies one is human insofar as they are part of the human lineage, rather than satisfying some necessary (set of) features that all and only humans have. His work undermined the normative value of a "type-specimen" by taking polymorphism and poly-typic representation seriously (1978, 351). The S-A-I thesis was not just theoretically significant, but socially conscious and non-exclusionary in ways that are sometimes overlooked.[10]

Next, evolutionary individuality is considered in its contemporary sense before other types of non-evolutionary biological individualities. Thereafter, the resulting plurality is analyzed.

---

[10] See Haber (2016) for a contemporary take on S-A-I.

Evolutionary Individuality

The Species-As-Individuals thesis was just discussed. Hull did not only focus on units of selection; evolutionary individuality had broader motivations. In contrast, recent work analyzes the nature of individuals *in selection* specifically. A main point of contention is to what extent reproduction, as an inheritance mechanism, is considered materially (i.e., in terms of material overlap and stability) or if formal interpretations of transmission are enough.

Below starts with a case study providing background from Lewontin (1970) and Janzen (1977). I close with a key takeaway about reproduction's conceptual expansion.

1. A Case Study & Some Background

Recall how Hull's (1976) individuality view used organisms as paradigm examples of individuality with integrated organization sustaining their unity. However, does integrative unity require parts to be in the same vicinity? Candidate criteria for organized unity include physiological integration (i.e., working together), spatial contiguity (i.e., being within proximity or in contact), and autonomy (i.e., sustained independence from external environment). However, cases lacking one or more of these attributes undercut these criteria (Santelices 1999, 152). Consider the following.

The Case of the "Humongous Fungus." In 1992, Smith et al. published an article in *Nature* claiming the largest and oldest living organisms were in the *Armillaria* genus. *Armillaria bulbosa* was identified as an individual occupying at least 15 hectares, weighing in excess of 10,000 kg, and retaining genetic stability for more than 1,500 years. *Armillaria gallica* extended up to 37 hectares of forest floor in Michigan's Upper Peninsula, but was older than the original estimates, at least 2,500 years old (see Anderson et al. 2018). Because asexually reproducing organisms occur across kingdoms in a variety of taxa, distinguishing asexually produced genotypes was essential for understanding their population biology to define their clonal structure. It (they? – that's the question) consists of nearly genetically identical fungi clusters often separated by trees, and in some cases entire forests. The clusters encompass tree root systems and exhibit stability of somatic mutations, which reflect historical growth patterns from a single point (2018). Why is this case a puzzle for individuality?[11]

In a 1977 landmark paper, evolutionary biologist Daniel Janzen argued that pre-theoretical intuitions misguide scientific work. Clonal cases – like the fungus above or dandelions, aspen groves, strawberries, and so on – are physiologically

---

[11] See Molter (2017) for analyses on the complexity of mushroom individuality.

distinct in terms of their location (i.e., the mushroom that can be picked, the dandelion for our wishes, the aspen tree cut down). Barring intra-clonal variation, Janzen argued that physiological unity is insufficient, and in some cases not even necessary, to distinguish individuals in natural selection. Genetic and phenotypic similarity exhibited by clones means they do not have varying traits populations of individuals require to evolve. In other words, they are not unique from one another from selection's point of view.[12]

Janzen's view motivates our fungi case: *Armillaria* in the upper peninsula of Michigan is one very large, spatially disparate evolutionary individual. It (no longer plural "they") defies pre-theoretical expectations due to its age as older than Christianity, its weight at 400 tons, and its size as larger than roughly seven Yankee stadiums put together (Pennisi 2018). Janzen's paper from 1977 mattered for challenging intuitions about individuality. He emphasized genetic identity and rejected the need for *physiological unity*: parts of an individual in selection need not be physically near and/or touching one another (i.e., spatially contiguous/continuous) to consider them *as parts* of that unit.

Evolutionary individuality, then, is where that theoretical machinery is put to use. Since some biological individuals are objects of natural selection, what's known about that process must be revisited. Lewontin's 1970 recipe is the received starting point.

According to Lewontin, three conditions must hold for evolution by selection. First individuals, the discrete units, must exhibit phenotypic variation. Second, those varying phenotypes must have corresponding rates of survival and reproduction, which indicate how varying traits make a fitness difference for the individual. Third, that individual's fitness, their ability to survive and reproduce, is based on the heredity of those traits; traits must be transmitted to offspring. And so, individuals in selection – evolutionary individuals – exhibit varying heritable traits that make a difference to their fitness. As Lewontin (1970, 7) puts it: "the primary focus of evolution by natural selection is the individual."[13] While the case above considered a clonal organism, as we'll see, not all evolutionary individuals are organismal.

---

[12] Clones are not unique from one another *other than their spatial location*, which could make a difference if fitness is partly determined by the environment.

[13] There are terminological issues concerning heritability vs. heredity/inheritance. Lewontin uses "heritability" in 1970. He later refined it in 1985. Godfrey-Smith (2009, 24) looks to equate "heredity" and "inheritance" distinguishing them from "herit*ability*": the former defined as parent-offspring similarity and processes concerning the inheritance of traits. The latter refers to a family of similarity measures determined by the statistical profile of the whole population (172). The present section centers on heredity, that is, how individuals form lineages and transmit traits, rather than the amount of influence those traits have on a population per se.

2. Darwinian Individuals, Reproduction, & Heredity

Hull (1980) and Godfrey-Smith (2009, 2013, 2015) build from Lewontin who did not analyze reproduction specifically. While Hull distinguished two types of individuals in selection (i.e., replicators and interactors), he focused on evolutionary theory's demand for historical entities. This included more than individuals in selection alone (i.e., units of mutation, of evolution). In contrast, Godfrey-Smith takes a narrower focus on individuals in selection, specifically the nature of reproduction as a mechanism for heredity.[14] Material overlap is too restrictive, and formal accounts of reproduction work well enough for Darwinian processes (2009, 83). Let's unpack this.

Godfrey-Smith (2009) draws from his view of Darwinian populations: Darwinian populations are evolving populations in which novel variations arise. Understanding reproduction as a way to satisfy heredity is an aim;[15] how individuals reproduce in those populations matter. 'Darwinian Individuals'[16] (his term for individuals in selection) need not be organismal: "genes, chromosomes, and other fragments of organisms can all form Darwinian populations" (2009, 85). Sometimes organisms don't meet evolutionary criteria, such as sterile animals who cannot produce fertile offspring. These organisms can metabolize but are unable to reproduce. They resist forces of decay, but cannot be individuals comprising Darwinian populations (2013, 25).

There are different forms or modes of reproduction. Simple reproducers, such as cells, make more of themselves by machinery *internal* to them. They often reproduce on their own in environmental contexts allowing for nutrients and energy to do that work. Scaffolding reproducers, like viruses and chromosomes, reproduce by means of structures or mechanisms *external* to them (2009, 88–89).

One mode receives special attention: collective reproducers have simple reproducers as parts, for example, human sexual reproduction and gamete fusion, but there can be collectives of collectives, such as eusocial insects containing multicellular organisms (2009). As a result of de-Darwinization, collectives form when evolution of lower-level entities is suppressed by evolution occurring at higher levels of organization. However, evolutionary activity at lower levels can disrupt stability of a collective unit, such as the rapid generation of cancer cells – simple reproducers – within a eukaryotic organism.

---

[14] Godfrey-Smith (2015, 10120) rejects Hull's replicator/interactor view siding with Lewontin. However, Hull (1980) also claims to build from Lewontin.

[15] Godfrey-Smith (2009, 25, 168) was also concerned with herit*ability*: tracking how a population responds to differences in fitness in population genetics.

[16] Evolutionary individuality is broader for Hull concerning historical (vs. ahistorical) entities, whereas 'Darwinian Individuality' for Godfrey-Smith concerns individuals in selection according to his view of population biology specifically.

Evolutionary processes at lower levels become *re*-Darwinized sometimes at the host's mortal expense.

Three parameters are identified that together set a gradient for collective reproducers: paradigm-minimal-marginal Darwinian individuality. How biological entities score on the following three criteria determines their place on that gradient.

First, Darwinian individuals exhibit degrees of physiological integration with the mutual dependence of parts serving different functions (2009, 93).

Second, an individual's parts exhibit a special division of labor, reproductive specialization, such as germline cells responsible for the capabilities of an organism to produce a new organism.

Third, there must be some mechanism by which the production of something new (i.e., re-production) is distinguished from growth of the same. Reproductive bottlenecks mark generational divides where genetic variation is reduced. In other words, reproductive bottlenecks force the process of growth and development to start anew (2009, 91). Consider how some organisms begin as one-celled zygotes that flourish to many cells: that narrowing is a "bottle neck" shape depicted in Figure 3 below.

What do puzzle cases look like on this view? Phenomenally individuated units of plants and fungi have narrowing runners that result in new clonal fragments. Godfrey-Smith (2009, 92) says, "considering again the cases with ramets and runners: the thinner the runner – especially in relation to what is to come – the less the new structure is a mere continuation of the old." This is a nonbinary view: "thinner" indicates a gradation from growth to the reproduction of something new.[17]

Something "new" is the reproduction of *offspring*, which matters if fitness is measured according to numbers of viable offspring.[18] Tracking parent–offspring relationships – lineages formed – is difficult in cases involving symbiotic relationships; associations with two or more partners from different species. Symbiotic associations vary in their nature, for example, how mutually beneficial the relationship is to all parties, whether there's continuous physical integration or containment, and so forth.

---

[17] PGS emphasizes gradation (2009), but Booth (2014) emphasizes types at the "poles": metabolic organisms and Darwinian Individuals. PGS defined reproduction along a gradient, however, metabolic organisms and Darwinian individuals diagram definite borders back onto categories he argued had a graded character. " ... for the purposes of summarizing basic relationships where some things occupy the middle space" (2009, 28–29). Continuums, gradation, and vagueness lend themselves to binary characterization. Conversely, there are points in a continuum even when the middle of the two poles is vague.

[18] Bouchard (2013) argues that sometimes fitness can be considered in terms of differential persistence, not reproductive success.

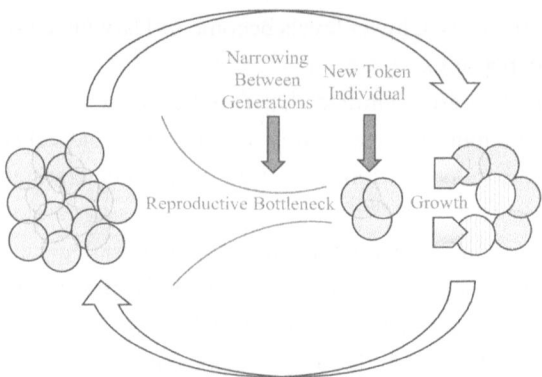

Figure 3 From many to few and back and again. New token individuals
distinguished by a narrowing reproductive bottleneck

On the one hand, Godfrey-Smith's view accommodates vertical transmission.
For example, aphid-*Buchnera* associations include bacteria that are transmitted
vertically through a lineage running in tandem with that of the aphid (2013, 31).
"Vertical" transmission concerns reproductive lineages of all symbiont partners
forming patterns of parent–offspring relationships arranged together – the
Buchnera are maternally passed to aphid offspring.

On the other hand, there's dispute over cases of symbiotic associations
recurring horizontally, for example, by uptake of partners from the environment. For example, while Hawaiian Bobtail Squid and its *Vibrio* bacteria form
a consortium, a potential adaptive unit, there is a many-many parent–offspring
relationship between symbionts creating complicated parent–offspring transmission networks.

So, microbial symbionts can be acquired vertically from host–parent to host–
offspring (e.g., like aphid-*Buchnera*), as well as horizontally from other host
organisms and the environment during development (e.g., like squid-*Vibrio*).
Godfrey-Smith (2015, 10123) argues that while the latter cases of horizontal
transmission result in multispecies metabolic collectives recurring through
actions of several Darwinian individuals, these are collaborations that coevolve,
but do not combine into single reproducing objects. That is, coevolutionary
theory accounts for those cases.

Godfrey-Smith's account of reproduction illustrates formal characteristics of
lineage formers and their degrees of "tightness" or obligation to one another in
parent–offspring networks. His concept of reproduction is satisfied "as long as
we know who came from whom, and roughly where one begins and another

Defining Reproduction for PGS

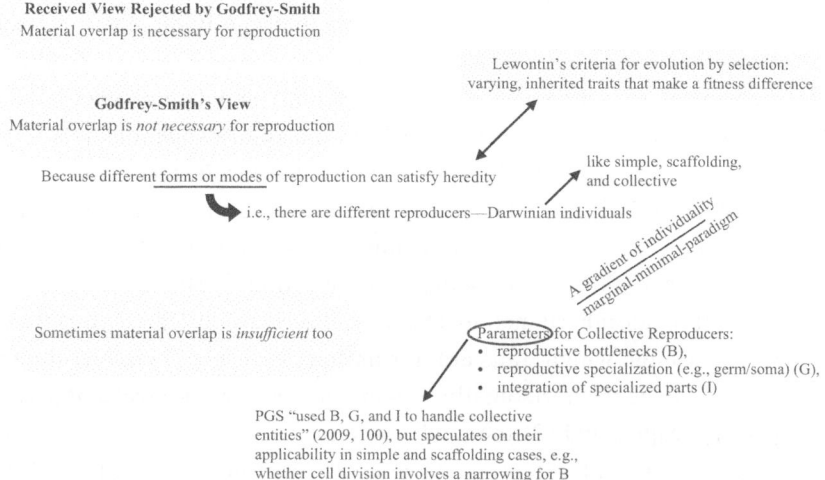

Figure 4 Material overlap is unnecessary for reproduction. Different forms of
reproduction satisfy heredity, that is, there are different reproducers –
Darwinian individuals – like simple, scaffolding, and collective, which each
scale along a marginal–minimal–paradigm gradient of individuality. Parameters
are set for collective reproducers: reproductive bottlenecks, reproductive
specialization (germ/soma), and integration of specialized parts. Sometimes
material overlap is insufficient too.

ends" (2009, 83). In that sense, passing on material structure is *unnecessary* (or
too restrictive) because there are different modes of trait transmission satisfying
heredity as summarized in Figure 4 above.

In some cases, like the squid-*Vibrio* example, material overlap is *insufficient*
for forming reproductive objects (i.e., Darwinian individuals) tracked to determine a population's response to fitness differences.

The complicated parent–offspring networks of horizontal transfer cases
confound fitness measures of reproductive success; viable offspring need to
be distinguished from parents (i.e., re-produced). However, one criticism is that
"the epistemological problems associated with tracking lineages of holobiont
parts should not lead to confusion about the fact that such patterns could be
determined *in principle*" (Booth 2014, 667). In other words, there may be
relevant ancestry, even if traditional concepts of parenthood are challenged by
complex (and reticulated) lineage patterns.

And while Godfrey-Smith argues that degrees of tightness matter for
heredity in multispecies units, others argue that inheritance can occur by
recruitment. Ereshefsky and Pedroso (2013, 2015) considered biofilms

candidate evolutionary individuals.[19] Biofilms are singular or multispecies communities of microorganisms; viscous collections existing in containers of water, streams, and rivers, and on living and nonliving surfaces. Cells exist together within an extracellular polymeric matrix, which facilitates both chemical signaling (i.e., quorum sensing) and lateral exchange of genetic material (i.e., by transformation or uptake of exogenous DNA, and by conjugation through "bridges" via plasmid DNA transfer).

Doolittle (2013, 372) said that *distinguishing* inheritance and recruitment matters for determining microbial communities like biofilms as bearers of adaptations, and thus as individuals in selection.[20] However, I specify recruitment *as* an inheritance mechanism. There are at least two different processes through which traits are transmitted, such as (1) via recruitment, which I take to be the horizontally convergent nodes within webs of parent–offspring lineages, and (2) the modes of reproduction that entail a vertical "tightness" of parent–offspring lineages as identified by Godfrey-Smith. However, while some satellite cells contain genetic material from the parent biofilm, it's debatable whether adaptive traits (e.g., like antibiotic resistance) exist at the level of the entire biofilm and are passed on.

Clarke (2016) objected that biofilms don't have the characteristics of lineage formers: they lack a unitary character because of aggregation, which means (1) they neither reliably pass on their structure to offspring biofilms (i.e., not sufficiently resembling them) nor (2) remain bearers of adaptations due to lack of stability. The frequent shuffling of genetic material makes it unlikely that "offspring" biofilms have traits resembling the parent biofilms.

In response, Pedroso (2017) maintained that overall function and corresponding phenotypes remain across successive generations. Aggregation yielding increased genetic variability through horizontal transfer does not mean that phenotypes themselves are unstable (e.g., such as antibiotic resistance, signaling patterns, potential for virulence, motility, etc.). Furthermore, biofilm recruitment is not an open-ended free-for-all. There are patterns of recruitment based on signaling – co-aggregation is a genetically controlled mechanism (129–130). Parent–offspring lineages are products of ecological succession. Their bottleneck patterns are formed by ecological surroundings, but that doesn't mean varying, inherited traits are absent (2017, 131).

---

[19] Following Ereshefsky and Pedroso, McConwell (2017a) argued that two types of individuals in selection are distinguished insofar as inheritance occurs in at least two different ways.

[20] Later Doolittle and Inkpen (2018) treat recruitment as something like reproduction, "reproduction", and thus, a sort of inheritance mechanism.

At this juncture, Doolittle and Booth's (2017) patterns of lineage formation are relevant. Functional patterns (e.g., metabolic and developmental interaction patterns), rather than the material (i.e., the taxa) responsible for them, are the units of selection. Material transfer is not needed for lineages formed by sequences of interaction patterns. While Godfrey-Smith emphasized modes of trait transmission, Doolittle and Booth focused on abstract functional relationships lineages engage with and present as their patterns through time.[21] Instead of reproduction, it becomes re*construction* – a formal kind of reproduction with no crucial piece of matter or material to be made (Doolittle and Booth 2017, 16, also see Doolittle and Inkpen 2018, 4007 on "re-production" meaning "created again"). While abandoning materiality, the abandonment of causality is denied. Patterns themselves are causal in the reconstruction of a new evolutionary individual.[22]

3. Inheritance Satisfied Functionally vs. Its Material Conditions

It is now time to draw a conclusion. Formal accounts of heredity emphasize functional profiles and patterns of inheritance. But there's varying attention to material conditions of trait transmission. Doolittle and Booth (2017, 6) say, "put metaphorically, what matters is the song, not the singer. The song, to flesh out the metaphor, is the pattern of interactions (metabolic, structural, or developmental) between partner lineages (the singers)." They argue that instances of interaction patterns pass on traits to later instances, leading to the differential persistence of the overall pattern's type. Those patterns, rather than material overlap and constitution, are sufficient to fulfill the criteria for evolutionary individuality. That is, heredity is fulfilled *formally,* and without appeal to inheritance's material conditions for making more individuals.

That "more-making" capacity, as Griesemer (2016, 807) puts it, has constraints though: "material overlap means that reproduction involves bonds of material continuity, not merely resemblance or formal information transmission." Griesemer (2005) previously argued that both information copying and formal relations are "problematic as stand-alone concepts of inheritance in abstraction from the material conditions of reproduction. The latter, not the former, determine the causal pathways of heredity relations" because the flow of genetic information depends on material connections between senders and receivers. Moreover, inheritance is a process where evolved mechanisms of

---

[21] Doolittle and Booth (2017, 16) add that that the patterns of interactional relationships lineages engage in are themselves a kind of replicator.

[22] Patterns are causal insofar as recurrences of metabolic pathways or interaction patterns are caused by previous instances of the same pathway or pattern. Avoid conflating: (1) the type of pattern and (2) its instantiation: a song in the abstract can be distinguished from its performance (Doolittle and Booth 2017, 19).

development are "propagated in reproduction," which must include both epigenetic and nongenetic mechanisms (2016, 807). Development is important because it is the "recursive acquisition, refinement, or maintenance of the capacity to reproduce" (2016). The driving point is to avoid considering evolutionary individuality in isolation from materials of developmental systems; it must be recast in research spaces of evo-devo, eco-devo, and developmental biology. There is too much focus on formal or overall structural patterns of inheritance, a critique best understood by the following example.

In a recent publication titled, "The Information Theory of Individuality" (Krakauer et al. 2020) the authors formalize how information is transmitted through an information-theoretic lens, while drawing from both environmental dependence and inheritance. Adaptive aggregations can be multi-scale without physical boundaries like cell walls or tissue, yet still visible to selection. They are evolutionary individuals in the sense discussed presently.

If information is only meaningful in the context of material systems, though, then the channel through which information is passed – the material medium – should not be overlooked. That is Griesemer's point.[23] Otherwise, what story is there about reproduction, if there is no medium between the sender and the receiver? Krakauer et al. (2020, 210) construe individuality without relying on material boundaries. They aim to capture fluidity and porousness of aggregates and associations, only the propagating forward of information though time is needed (214). Quantity, closure, and autonomy are defined by formal boundaries of functionally individuated systems emerging at different scales (220). *Not* required on their view is the kind of physical contact, overlap, and raw materials involved in causal pathways of heredity relationships.

The key takeaway is that evolutionary individuality facilitated conceptual expansions of reproduction into more abstract or formal (versus material) senses. Godfrey-Smith (2009, 84) argued that nothing about inner logics of Darwinism preclude a purely formal account. Others proposed inheritance by recruitment; re*construction* of new offspring individuals distinguished by ecological bottlenecks, which comprise parent–offspring lineages. The functional shape of reproductive patterns was taken further; *transmission* patterns – their information – transcend physical boundaries yet cause variation remaining visible in *in its form* to selection (see Figure 5).

However, pre-occupations with purely formal aspects abstract from reproduction's material conditions, which risks occluding matter – the medium – that both determines the causal pathways of heredity relations

---

[23] Griesemer (2016, 807) says, "An issue that divides me and Godfrey-Smith is what counts as a salient material bond." The disagreement turns on, for example, formal/informational relations in retroviral replication versus material overlap due to RNA strand hybridization channeling genetic information.

Debate Over Lewontin's Heredity Condition: Formal Modes & Material
Conditions of Reproduction

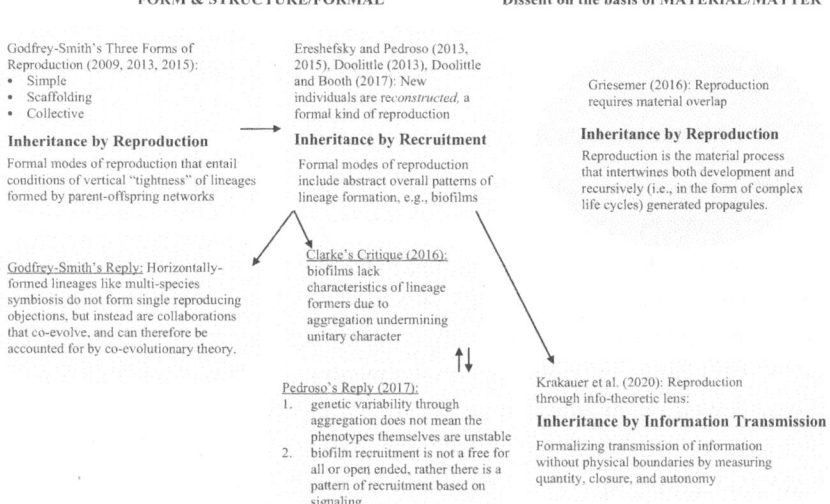

FORM & STRUCTURE/FORMAL   **Dissent on the basis of MATERIAL/MATTER**

Godfrey-Smith's Three Forms of
Reproduction (2009, 2013, 2015):
- Simple
- Scaffolding
- Collective

Inheritance by Reproduction

Formal modes of reproduction that entail
conditions of vertical "tightness" of lineages
formed by parent-offspring networks

Ereshefsky and Pedroso (2013,
2015), Doolittle (2013), Doolittle
and Booth (2017): New
individuals are *reconstructed*, a
formal kind of reproduction

Inheritance by Recruitment

Formal modes of reproduction
include abstract overall patterns of
lineage formation, e.g., biofilms

Griesemer (2016): Reproduction
requires material overlap

Inheritance by Reproduction

Reproduction is the material process
that intertwines both development and
recursively (i.e., in the form of complex
life cycles) generated propagules.

Godfrey-Smith's Reply: Horizontally-
formed lineages like multi-species
symbiosis do not form single reproducing
objections, but instead are collaborations
that co-evolve, and can therefore be
accounted for by co-evolutionary theory.

Clarke's Critique (2016):
biofilms lack
characteristics of lineage
formers due to
aggregation undermining
unitary character

Pedroso's Reply (2017):
1. genetic variability through
   aggregation does not mean the
   phenotypes themselves are unstable
2. biofilm recruitment is not a free for
   all or open ended, rather there is a
   pattern of recruitment based on
   signaling

Krakauer et al. (2020): Reproduction
through info-theoretic lens:

Inheritance by Information Transmission

Formalizing transmission of information
without physical boundaries by measuring
quantity, closure, and autonomy

Figure 5 Schematic of debate over Lewontin's heredity condition: formal
modes and material conditions of reproduction. On the left, reproduction is
formally analyzed by abstract functional relationships lineage-formers engage
in. On the right, there is dissent from a material standpoint targeting abstract,
formal accounts. Insofar as evolutionary individuals must pass on varying,
inherited traits, how that occurs is analyzed both formally and materially

and the effects of development on reproductive systems (Griesemer 2016,
201). Reproduction's material processes intertwining both development and
recursively (i.e., in the form of complex life cycles) generated propagules
must not be overlooked.

Relevant material conditions could be further analyzed by drawing from Lynn
Margulis' (e.g., 1998) work. Margulis argued that symbiogenesis – the merging
and diverging of associations – drives evolutionary change. For Margulis, cell
structure, biochemistry, and geological context are material conditions that matter
for increasingly complex levels of individuality. New adaptations arise not only
from random mutations but also the merging of two separate organisms in those
conditions (see Goldscheider 2009, 44). This is a task for later work.

To summarize, reproduction as a heredity mechanism for individuals in
selection was analyzed according to both formal and material accounts. Next,
individuals in immunology, ecology, and metabolism are considered.

Immunology, Ecology, & Metabolism

Biological individuality was just discussed within the context of evolution. Going beyond evolution, immunological, ecological, and metabolic accounts of biological individuality are considered below. Threads of life and health become salient here.

1. Immunological Individuality

In a 1992 chapter, David Hull stated that if physiology were theoretically developed enough, it could be used to determine individuality in biology due to its emphasis on mechanics and function of multi-part systems and wholes. Pradeu (2010, 2012) takes up this challenge. Functionally integrated parts are pervasive in literature on biological individuality, even of the evolutionary sort: Godfrey-Smith's Darwinian individuals must be integrated and exhibit relevant divisions of labor among parts that support their mutual dependence. Hull (1978) argued that functional integration is a necessary feature. Providing specificity to what "integration" means, Pradeu draws from physiology, in particular the field of immunology.[24]

Roughly, immunology is the study of how bodily systems fend off threats, but how to define immune response is contentious. For Pradeu, immunogenicity, or the triggering of immune response, occurs in the presence of strong discontinuous molecular difference rather than exogenecity specifically, that is, rather than a foreign "non-self" source (137). Pradeu (2012, 143–144) specifies that immune response does not concern just any discontinuity, but a strong discontinuity of ligands with which immune cells interact. Relevant factors of discontinuity include features of antigens or substances that induce immune response, such as antigen quantities, speed of antigen appearance, degree of molecular difference, and regularity of antigen presentation.

A referee requested that because molecular interactions are foundational to Pradeu's view, a note on genetics should be provided. However, it is imperative not to conflate genetics-based analyses with physiological study of molecules and receptors in, on, and around the surfaces of (immune) cells.[25] Genetics is not all there is to molecular activity – cellular and molecular physiology concern biochemical interactions and dynamics. Insofar as genetics concerns the study of inherited differences in evolutionary contexts, a main framing of Pradeu's book is Hull's challenge to provide a well-developed physiological, rather than

---

[24] Pradeu (2012) offers the Continuity Theory of immunology, to avoid problems that Burnet's Self/Nonself Theory couldn't overcome, such as non-threatening responses to endogenous or "self-sourced" tumor antigens and tolerance of exogenous or "nonself-sourced" symbiotic bacteria. Historically, immune response was defined by presence of foreign, genetically different, "nonself entities" (see Pradeu 2012, 57). Pradeu argues that strong modification of antigenic patterns is what elicits immune response.

[25] See Pradeu (2012, 139) for a list of relevant immune receptors.

evolutionary, theory of individuality and identity in biology. Moreover, Pradeu distinguishes between the continuity theory at the genetic level versus its application concerning molecular patterns recognized by receptors mostly located on the surface of immune cells (see 2012, 178).

In contrast to previous discussions of evolutionary individuality, Pradeu's work highlights *maintenance conditions*, rather than reproductive conditions for new token individuals. For example, if cancer cases constitute breakdowns of cohesive functioning with parts proliferating at the expense of the whole, then Pradeu's immunological account explains when and how individuality fails: one-way individuality breaks down is through poor boundary maintenance and control over parts. With new technologies in cancer therapeutics drawing from immune response research, close attention ought to be paid to immunological individuality.

Considering an evolutionary context Pradeu centers organismality; "it is necessary to examine the physiological processes produced in the organism to arrive at a precise definition of what, in each case, counts as an evolutionary individual" (2012, 260). While organisms are not always individuals in selection, starting with the heterogeneous organism helps to determine what evolutionary individuals *are*.[26] An immunological account identifies organisms as the most well-defined individuals, rather than merely one individual among other types (2012, 264).

For a historical take, consider Medawar (1957), a zoologist and comparative anatomist, who investigated the uniqueness of individuality. He explored issues with surgical skin grafting and transplantation, that is, why intolerance occurs when borrowing from members of one's own species in humans. Medawar also considered immunological reactions (e.g., to bacteria and viruses, but also allergens). He disagreed with philosophers that the distinction among individuals is of a difference in kind or even of degree (1957, 154). Instead, he proposed that difference among individuals amounts to something combinatorial: one individual differs from others not because of unique endowments, but because of the unique *combination* of endowments (1957). Medawar investigated how combinatorial factors of one individual are retained, for example, what the antigens are and what keeps the immunological reaction maintained. Indeed, immunological individuality is intertwined with medical consequences.[27]

---

[26] Pradeu (2012, 177–180) applies continuity theory to plants and unicellular organisms. Additionally, fetomaternal tolerance and chimerism (i.e., the pregnant person tolerating and conserving cells from a carried fetus) are no longer isolated exceptions to self/non-self differentiation. Instead, Pradeu (pp. 116–117) accounts for situations of tolerance in pregnancy, as well as in commensal and symbiotic micro-organismal associations existing in the digestive tract and epithelia. This occurs by way of an organism's active increase of tolerance through its repertoire of non-immunogenic antigens.

[27] While Julian Huxley argued for a progressive version of individuality with humanity at the pinnacle (and scientific management in a eugenics sense to "improve" it), Medawar (1957, 185) says, "So far from being one of his higher or nobler qualities, his individuality shows man nearer

2. Ecological Individuality

Are communities in ecology mere assemblages? Or individuals in their own right? Lean (2018, 520) argued that ecological communities are often *not* biological individuals because they lack causal boundaries – individuality is not the natural end point of all biological interactions.[28] Others maintain that ecological communities are individuals in contrast to arbitrary sets of things (e.g., see Huneman 2014a and 2014b). Complex interactions within communities are "stronger" than interactions between communities and the external environment. Huneman explores different theoretically driven definitions of "strength" because "within a theoretical domain [they allow] us to partition the assemblies into "individuals" and "non-individuals" (2014, 361). Ecological individuality is a continuum: strong and weak individuals are distinguished by the connections among their constituents. Interactions exist within a set of formally defined parameters in terms of how likely they are to occur from more intimate interactions (e.g., aphid-*Buchnera* consortia) to interactions with lower intimacy (2014, 370). Engineering ecosystem interactions involve several organisms from many species over a significant timescale. When considering ecosystems, the living or "biotic" criterion appears less stringent.

Ecosystems include organismal interactions with *abiotic* components like the surrounding landscape (e.g., soil, water, and so forth). Millstein (2018, 281) explores the concept of a land community as articulated by Aldo Leopold, a twentieth-century forester who, as Millstein argues, "did seem to think that the land community was an individual": he didn't use the word 'individual' specifically, though he did explore whether the land community is an organism.[29] Since 'organism' and 'individual' were often used synonymously at the time, Leopold was likely probing a land community's individuality status by exploring its organism-potential. Leopold spoke of interdependence among "soils, waters, plants, and animals *collectively*" and considered both the organization (and sometimes disorganization) of the land (2018, 281). What's at stake for the land community is its status for moral considerability, that is, whether it can be an object of moral obligation and have intrinsic value. The moral motivation canvased by Millstein is a normative spin on individuality in the

---

kin to mice and goldfish than to angels; it is not his individuality but only his awareness that sets man apart."

[28] Instead, Lean (2018) maintains that ecological communities are units indexed against the network of weak interactions between populations that unfold from a starting set. He challenges the false dichotomy between mere assemblage and individuality by exploring ecological complexity through a different lens. Individuality is not the only organizational principle of nature.

[29] Millstein (2018, 204) clarifies that Leopold's land community concept combines aspects of "ecological community" with only biotic components, with an ecosystem approach that includes abiotic components.

biological domain. However, the individuality status of land communities is difficult to establish.

Distinguishing boundaries is complex because open systems, that is, "systems where the spatial area of the densely interacting populations is larger than that of the dense matter/energy flow – or vice versa," challenge traditional conceptions of boundaries as strictly physical or material barriers (2018, 291). While sometimes the spatial contiguity and continuity of an individual's parts determine boundaries, in the land community case Millstein considers (and rejects) congruence as a requirement: parts need not be the same, nor need they always "agree" or be compatible in way that is strictly cooperative. As she puts it: "Individuality does not require location in the same space": parts can be spatially disparate. Systems can be well-bounded or open, but still satisfy the metaphysical constraints as identified by Ghiselin (1974) and Hull (1976), such as restricted to time and place, integrated parts such that the mutual dependence of causal interactions affects their shared fate, beginnings and endings in time, and continuity through time (Millstein 2018, 297). The problem, as Millstein (298) identifies it, is distinguishing an individual from "an abstract type or a mere set or a mere assemblage," of which "organismality, internal regulation, being a unit of selection, and/or emergent properties" are not necessary for it.

Millstein points to features like internal regulation that could make an individual more robust but would not be necessary. While "robustness" carries connotations of resilience, in the context of ecological individuality, resilience against what? One might consider what defines individuality within a particular domain, such as an ecological individual's ability to persist under various perturbations like pollution, poaching, and invasive species. In that sense, resilience is normatively cast as ecological *health*. Developing a concept of health according to both ecological and immunological individuality indicates the need to coordinate those domains. If different domains interact in important ways, overlap should be expected:

> Ecology is the study of distribution and abundance of organisms and their interactions with their environment, including parasites and pathogens. Immunology is the study of the physiological functioning of the immune system in states if health and disease. The former discipline [ecology] acknowledges the importance of the latter [immunology] but treats it as a black box (Schulenburg et al. 2009, 3).

Thus, resilience produces robust individuality, which matters as a *normative* feature, one that arises in both immunological and ecological contexts concerning resilience against disease, poor function, and breakdown. Addressing heterogeneity and boundary conditions certainly matters, but healthy functioning in these contexts, however defined, reveals normative concerns about how ecological and immunological individualities *should* be.

In other words, while Millstein posits moral considerability as one normative element at stake, healthiness of ecological and immunological individuals is another. This is one angle for approaching the ecology of holobionts: microbes and their host-organism relationships occur in networks of interactions and dependence. Gilbert and Tauber (2012) argue that eco-immunology provides evidence for a holobiont's individuality, which includes surveillance and response immune mechanisms. Those mechanisms play a critical role in regulating healthy social ecologies of holobionts, while acknowledging microbial activity in the immune system itself (2016, 846–847). Immunology and ecology are not all there is though. Threads of life and health arise in another context for biological individuality.

3. Metabolic Individuality

What are metabolic individuals? They might just be organisms, at least for Godfrey-Smith (2009) who distinguished between metabolic organisms and reproducing Darwinian individuals. Some organisms cannot reproduce (e.g., sterile mules or castes of insects). If organisms are defined immunologically or even ecologically, why bother discussing metabolism within the context of biological individuality at all? While Dupré and O'Malley (2009) bring metabolism and replication/reproduction into the context of biological individuality, the issue is framed within the context of life.

Life is a complex concept with debate over necessity of metabolic function to consider something alive. For example, viruses may replicate themselves under the right conditions, but are metabolically inert and so are not considered alive (unless we focus on replication as a sole criterion for life). If one assumes that biological individuals must at minimum *be alive* (though the abiotic components of land communities considered previously might suggest otherwise), then how life is defined impacts what's included under biological individuality's scope. However, Godfrey-Smith (2016) points out that life has two sides: a metabolic side and another side concerning reproduction and evolution. To ask whether one is primary or sufficient is a misguided question. And so, we might consider viruses as biological individuals of a particular sort, even though they do not really have their own metabolism: they form Darwinian populations as they engage external machinery to make more copies of themselves (see Godfrey-Smith on scaffolding reproducers).

But let's stay focused on biological individuality and metabolism. Metabolism amounts to conversion: the chemical processes and reactions involved in sustaining energy production, and the breakdown and synthesis of molecules and compounds needed to resist forces of decay. If considering individuality at a metabolic scale, then biochemical composition must be measured and tracked.

Beebe and Kennedy (2016) describe technology to comprehensively measure and track an individual's metabolic profile, that is, metabolomics. Contrast metabolomics with genomics; "metabolomics" refers to the metabolic profile (i.e., the character or quality of an organism's metabolic activity), rather than genomic profile, with the purpose of refining therapeutic interventions in precision medicine. Beebe and Kennedy offer a means of drafting blueprints – technical maps of molecular underpinnings of human individuals beyond genetics.[30] Philosophers might recognize blueprinting as drafting the identity of metabolic individuality: what makes one individual unique from another in a biochemical sense.

However, organic individuation through the lens of metabolic reactions has been around at least for a while, such as C. M. Child's *Individuality in Organisms* (1915) where he develops a dynamic conception of individuality such that maintenance is possible in a changing environment. To be clear, Child's solution takes a physio-morphological spin as he wrestles with concepts of unity and order. He searches for physiological evidence of metabolic gradients: The organic individual is fundamentally a dynamic relation of dominance and subordination, associated with and resulting from the establishment of a metabolic gradient or gradients (1915, 88).

An organism's "axis" (e.g., analogous to Earth's axis – the imaginary pole running from "top" to "bottom") represents direction of a gradient from higher to lower rates of metabolic reactions, such that, if I'm reading Child correctly, those rates decrease in the direction of that axis (1915, 20). Child was dedicated to finding evidence for dynamic boundaries of metabolic individuality.

Taken together, individuality in immunology, ecology, and metabolomics (metabology?) is a varied and interconnected landscape relevant to life and health. While much philosophical attention has directed toward individuals in evolutionary biology, these are solid foundations for branching to other non-evolutionary biological sciences.[31]

Up to this point, biological individuality was approached from multiple vantage points including evolutionary individuality (in broader and more specific senses concerning individuals in selection), and different types of non-evolutionary

---

[30] Beebe and Kennedy (2016, 99) argue that genetics alone is insufficient to explain phenotypic traits relevant in medicine.

[31] Unfortunately, due to space constraints this section lacks developmental individuality, a strong current in the third section of this Element and included in the next section's Table 1.

biological individuals. Next, some conclusions are drawn before turning toward critical discussion of biological individuality in the production of scientific knowledge.

Pluralism, Historical Hierarchy, & Ambiguity

It is time to take stock of the ontic landscape developed so far. Consider Table 1 on views concerning biological individuality below.

I call these approaches "domain-driven" because their analyses derive from select disciplinary domains or subspecialities. The table organizes views into a conceptual morphospace, which is *not* historical or developmental but instead identifies subject-led entry points for tractability in a complex interdisciplinary landscape. How is this plurality to be understood?

As we've seen, any one type of biological individuality includes a variety of entities. Examples of evolutionary individuals range from organisms to viruses to genes, while metabolic individuality will include organisms and exclude viruses. There is some cross-classification: many evolutionary individuals can be both immunological and metabolically defined, but some organisms lack features to be objects of evolutionary processes. While a shared research aim, 'pluralism' is often used ambiguously. I provide critical points of discussion in that regard before drawing a more positive picture of pluralism both at a time (i.e., synchronic) and over time (i.e., diachronic).

1. A Critique: Heterogeneity & Pluralism

First, recall how Godfrey-Smith (2009) argued that individuals in selection reproduce by yielding reproductive bottlenecks, exhibiting reproductive division of labor, and behaving as functionally integrated units. He recognizes degrees of individuality – entities score high on these parameters as paradigm individuals versus marginal individuals that score low on all three. Booth (2014, 671) views the account as pluralistic because Godfrey-Smith identifies two types of biological entities – metabolic organisms and Darwinian individuals. This is a plurality of biological individuals, but not of individuals in selection specifically. Godfrey-Smith's definition of individuals in natural selection is constrained by a single set of parameters where entities fall along a gradient. Different entities, for example, viruses and organisms, can become or cease to be Darwinian in the evolutionary sense.

Table 1 Individuality views

Subject	Author	Disciplinary Focus	View
Organismality	Huxley (1912) *Individual in the Animal Kingdom*	Evolutionary biology and eugenics	Organisms are representatives of individuality. Living matter groups itself into units, never complete boundary closure, independence/autonomy is never absolute, and harmony of parts never perfect.
	Sir Osler (1904) Ingersoll Lecture	Biomedical Science	Organismality is fleeting and ephemeral; ontologically secondary to eternal generational threads or "embryonic substance."
	Ruse (1987)	Philosophy of biology	Organisms are decomposed into structurally-various parts functioning together interdependently to sustain the whole unit (also see Ghiselin 1987; Hull 1978).
	Nuño de la Rosa (2010)	Philosophy of biology, evo devo	Organisms are functionally-integrated and autonomous systems, have a larger theoretical role besides evolutionary theory. A holistic view emphasizing development and activity (also see Huneman 2017).
Individuals in Evolution	Ghiselin (1966)	Biology, philosophy, history of biology	Species must be individuals to change and evolve (vs. classes/kinds as static with unchanging sets of necessary conditions).
	Hull (1976, 1978, 1980, 1992)	Philosophy of biology, history of biology	Metaphysical individuals are discrete, unique, continuous in time and space, and unified by functionally integrated parts. Species and organisms fit this definition (also see Ruse 1987; Haber 2016). Hull identifies units of mutation, of natural selection (i.e., replicators

Table 1 (cont.)

Subject	Author	Disciplinary Focus	View
Individuals in Selection	Janzen (1977)	Evolutionary ecology and conservation	and interactors), and of evolution. Physiological unity, contiguity are insufficient for individuals in evolution, genetic identity matters. Units of selection can be spatially disparate. Also see Lewontin's (1970) primary object of evolution by selection as the individual.
	Godfrey-Smith (2009, 2013, 2015)	Philosophy of biology, population biology	Darwinian Individuals comprise evolving populations, are the bearers of varying, inherited traits that make a fitness difference. There are different forms of reproduction or "reproducers" (i.e., simple, scaffolding, and collective), which reproduce according to three parameters comprising a gradient.
	Ereshefsky and Pedroso (2013, 2015), and Pedroso (2017)	Philosophy of biology	Drawing from Hull. Replicators pass on their structure largely intact (e.g., genes), and interactors cohesively interact with their environment (e.g., like organisms) with unitary effect on constituent replicators. Multi-species biofilms are evolutionary individuals despite scoring low on Godfrey-Smith's criteria. Also see Clarke (2016) for critique on biofilms as adaptation bearers, McConwell (2017) for two kinds of evolutionary individualities, and Doolittle (2013) for inheritance by recruitment.

Table 1 (cont.)

Subject	Author	Disciplinary Focus	View
	Krakauer et al. (2020)	Evolution of information processing, theoretical computer science	Adaptive aggregations are multi-scale without physical boundaries, yet still visible to selection. See Doolittle and Booth (2017) on patterns of interactions between partner lineages as visible to selection. See Griesemer's (2005, 2016) contrasting emphasis on materiality.
Individuals in Immunology	Pradeu (2010, 2012, 2016)	Philosophy of immunology, biology	An immunological account identifying organisms as most well-defined individuals, rather than merely one among many types.
	Medawar (1957)	Zoology, comparative anatomy	Uniqueness not in terms of kind or degree, but a combinatorial difference.
Individuals in Ecology	Huneman (2014)	Philosophy of ecology and biology	Ecological communities are not arbitrary assemblies. Interaction strength formally-defined distinguishes strong/weak individuals (Also see Lynn Margulis on symbiogenesis, Lean 2018)
	Millstein (2018)	Philosophy of biology, ecology	Land communities with both biotic and abiotic components. Systems can be well-bounded or open still satisfying constraints of locality in time and place, interaction, and continuity through time, as well as a shared fate.
Metabolic Individuality	C. M. Child (1915)	Zoology, regeneration	Physiological evidence of metabolic gradients accounts for dynamic boundary maintenance. See Beebe and Kennedy (2016) on measuring a metabolic

Table 1 (cont.)

Subject	Author	Disciplinary Focus	View
			profile of individuals to tailor therapies in precision medicine.
	Dupré and O'Malley (2009)	Philosophy of biology	Debate over metabolic function as necessary for alive-ness or replicative capacity alone. If biological individuals are minimally *alive,* then views about life impact views about biological individuality. See Godfrey-Smith (2016, 2009): metabolic organisms vs. Darwinian individuals.
Individuals in Development	Maienschein (2011)	History of biology, developmental biology, and embryology	Organismal individuality is considered by what constrains and enables development. The history of biology shows how the organization of regulatory wholes defines the boundaries of development.
	Griesemer (2018)	Philosophy of biology, developmental biology, evo-devo	The developmental individual is a reproductive process marked from start to finish, transitions between reproductive generations are defined by developmental stages. The organism is a tracking tool for delimiting reproductive generations of individuals from developmental ones.

Second, recall Huneman's (2014a, b) proposal that strength of inter-action marks boundaries of individuals. Evolutionary individuals exhibit strongest levels of interaction delineating interactions of entities with one another versus with entities outside of that unit (2014b, 377). Increasing levels of interaction produce individuality pluralism because nested individualities comprise levels of biological hierarchy (374). Different entities across scales of interaction, in an ecological sense, also become or cease to be evolutionary individuals depending on the strength of interaction among them. In other words, evolutionary individuality as one type of biological individuality is distinguished from ecological individuality (i.e., another type of biological individuality) by way of interaction strength.

To clarify, Godfrey-Smith and Huneman use 'pluralism' to describe their views concerning many types of biological individuals. There are at least three ways to consider pluralism and biological individuality:

(1) Many types of biological individuals that are roughly domain-specific (e.g., evolutionary, metabolic, ecological).
(2) Many types of evolutionary individuals, of ecological individuals, of metabolic individuals, and so on (i.e., each type has a plurality of subtypes).
(3) Both (1) and (2).

In (1), biological individuality is a heterogenous category of all types of biological individuals. Those types are defined by different concepts, concepts which range over complexes of processes and patterns relevant for (and across) subdisciplines of the life sciences. Several classificatory concepts characterizing biological individuality differently (evolutionary, ecological, immunological, and so on) were previously discussed.

However, when unpacking the prospect of (2) (and thereby (3)), consider how several classificatory concepts characterize, for example, *evolutionary* individuality differently.[32] That is, concepts that organize the class or category of all individuals in selection, what I call the 'EI category'.[33] Let me explain.

Taking Lewontin's recipe to define the class of all individuals in selection, our EI category, then at minimum all characterizations of EIs must exhibit inherited variation in fitness. However, there are several views on how individuals satisfy Lewontin's heredity constraint: through formal senses of reproduction to transmit traits (Godfrey-Smith 2009, 2013, 2015), of *reconstruction* including non-vertical forms of transmission of adaptive traits (Ereshefsky and Pedroso 2013, 2015; Clarke 2016; Pedroso 2017; McConwell 2017a, b), and of transmitted interactional patterns in a functional sense with differential persistence (Doolittle and Booth 2017). If heredity is satisfied in different ways

---

[32] Mutatis mutandis for other types: whether there are multiple types of immunological individuals, or ecological individuals is a task for another day. I focus on the contemporary sense of evolutionary individuality, namely, individuals in selection.

[33] See Havstad (2021, 7671–7672) for ingredients to disentangle elements of classification: characterization, individuation, and organization.

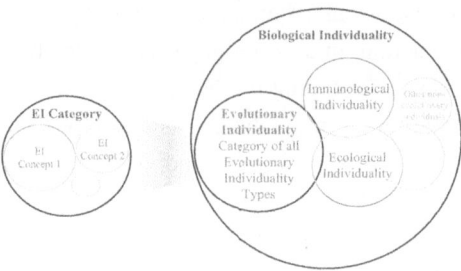

Figure 6 EI category vs. EI concepts. The EI category is defined by Lewontin's view of evolution by selection. EI concepts organize candidates for evolutionary individuality in different ways. To be a pluralist about EI concepts is to reject that there is one single correct way to define the EI category

under Lewontin's view (in the sense of McConwell 2017a, b), and each way generates a different EI concept, then candidates for the EI category are organized according to how heredity is characterized.[34] Consider Figure 6.

The conceptual typology of EI is organized further according to functional and material approaches. For example, if the typology is defined functionally, then heredity for Lewontin might just be one constraint on the functional or theoretical role of evolutionary individuals realized multiple ways. That aligns well with Clarke's 2013 view: evolutionary individuality is a theoretical role realized by varieties of (reproductive) mechanisms. However, if the mechanical details of inheritance matter for pluralism, a plurality of evolutionary individuality *concepts* can be distinguished on the material basis of how trait transmission occurs. This pluralism is all about perspective.

There would be dispute over the EI category, the class of all evolutionary individuality types, if there were dispute over how evolution by selection works. So far, there seem to be variations consistent with the original Lewontin formulation; slightly different accounts of the "recipe," such as Okasha (2006), Brandon (1999), and Bouratt (2015). Hull's replicator-interactor account may be just an extension of the original Lewontin formulation. The Price Equation might be as well. Okasha (2006) explores the link between Price's equation and Lewontin's account of evolution by selection. Price's equation tracks combined effects of two or more levels of selection on evolutionary change (2006, 18ff). The point is that a different recipe would characterize the EI category differently and as such shift the basis of present discussion.

While a synchronic typology of biological individuality is ontically oriented, disciplinary dimensions of Table 1 further muddy the picture. I return to this in

---

[34] Or similarity, depending on how material conditions of inheritance are characterized.

a moment. For now, disambiguating pluralism(s) in part depends on what sort of pluralism one has in mind.

2. A Positive Proposal: Synchronic & Diachronic Plurality in Life's History

How do new individuality types emerge over time? I propose shifting toward biological individuality's broader context: the history of life. The following considers how *new types* of individuality occur over time, a diachronic view.

There is a strong connection between individuality and major transitions in evolution. Major transitions mark pivotal turning points in life's history (Maynard-Smith and Szathmary 1997). Some evolutionary events are deemed major, like transitions from unicelled to multicelled life, because of momentous shifts in organization. Life's history and hierarchy contextualize individuality pluralism against timescales. How things come together or "associate" matters for that story.

Symbiotic associations occur (1) between two or more organisms from different species and (2) are constituted by varieties of mechanisms. In endosymbiosis, physical merging events occur with one partner subsumed within boundaries of another. In ectosymbiosis, the partner remains outside, but on the surface. In mutual associations, the relationship is beneficial to all parties. Commensalism occurs when only one partner benefits with no positive or negative benefits to the other. Parasitism occurs when one partner receives positive benefits and the other negative benefits.

Symbiosis, particularly endosymbiosis, contributes to the emergence of new individuals through a union. For example, a well-known origin story of eukaryotes outlines how one type of cell, an archaean, engulfed a proteobacterium over 1.5 billion years ago. That proteobacterium evolved into what is now known as mitochondria (O'Malley 2014, 29). This case is a transition: symbiosis generates new individualities. That's unsurprising to someone like Lynn Margulis (1998) who argued in favor of symbiogenesis theory. Symbiotic relationships are sources of innovation. New individualities are formed by means of incorporation (rather than competition).

Another way associations occur is through aggregations of cells. At what point is an aggregate one individual rather than mere assembly? Hammerschmidt et al. (2014) discuss the role of cooperation, cheater suppression mechanisms, and controlled generation of cheater phenotypes using *Pseudomona fluorescens*, their model organism. Rainey and Kerr (2010) approach the same case study but in reference to life cycles, collective reproduction, and again cheater control. Others study how collectives of *Saccharomyces cerevisiae* form a multicellular collective in response to environmental stress that includes evidence of reproductive division of labour (e.g., Libby et al. 2014). And finally, additional model organisms are used to investigate transitions to multicellularity, including unicellular and multicellular phases in life cycles, such as *Dictyostelium discoideum* (slime molds) or other social amoeba (Queller and Strassmann, 2012). Studying transitions to multicellularity, these researchers explore how new types of (often evolutionary) individuals emerge with features across levels of organization comprised by entities previously units of their own. The microbial world changes how evolutionary processes are viewed (e.g., see Sapp 2009).

Transitions from mere assemblages to individuals sometimes occur among closely related entities, like daughter cells glued together in bacteria and multicellular yeast; *P. fluorescens* and *S. cerevisiae*. Other times transitions occur among entities not closely related, such as multispecies biofilms, or our endosymbiotic origin story of eukaryotic mitochondria mentioned above. Queller (2000) identifies the former "closely related" transitions as fraternal and the latter as egalitarian where allegiances are forged such that "higher-level" units emerge.

While major transitions in the history of life involve more than transitions in individuality (e.g., oxygenation of the earth, see O'Malley 2014), Evolutionary Transitions in Individuality, ETIs, mark some major events (Michod and Roze 2000). ETIs are caused by evolutionary processes, and their occurrence affects evolutionary processes. Individuality's role in Leo Buss's *The Evolution of Individuality* (1987), as the title suggests, is integral to the evolution of hierarchical organization. He views life's history as a "history of transitions between different units of selection (1987, 171). Put differently, individuality is the answer to the "why" when Buss asks, "why is life hierarchical?" (183). Individuality transitions are diachronically productive of new types. But can individuality ever be destructive?

While much work focuses on transitions to multi-celled life as examples of ETIs, consider the prospect of a *failed* transition. Godfrey-Smith discusses the process of de-Darwinization: when lower-level competition suppression, cheater controls, and so on fail such that selection drops a level, often to the detriment of the higher-level unit. Cancer is case and point: cancers as diseases of multicelled life include out-of-control cellular proliferation in a variety of forms (see Plutynski 2018). Is cancer the emergence of new (or old?) individuals? Or is it the breakdown of multicelled individuality? As a breakdown, cancer could mark a *major transition failure*. Transitions, such as transitions in individuality, are often considered complete and in the past, but some transitions are still in progress and success is not guaranteed. Whether the emergence of new individualities necessarily destroys old types is beyond the scope of present discussion.

The key issue is that individuality according to *historical timescales* shifts the vantage point. The emergence of new individualities supports a picture of evolutionary individuality pluralism over time (McConwell 2017a, b).[35] How much change is necessary to distinguish a new type of individuality, or whether forms of individuality in some sense "speciate", will depend on:

(1) how we think about evolution, and
(2) the constraints of our typology (i.e., how types are defined in a system of analysis).

That is, evolutionary individuality *as* an evolutionary product takes a diachronic plurality of forms: different types of individualities likely emerge, evolve, and disappear over the course of evolutionary time (see McConwell 2017b, 131ff for a view on diachronicity and synchronicity of types, and Currie 2019b, 38ff on the emergence of different kinds through historical processes). Diachronicity requires some flexibility in how we think of evolution itself: is evolution a stable process? Or do evolution's causal processes evolve? Some have looked for evidence of, for example, the evolution of heredity mechanisms themselves (as per Darlington 1958). Others have considered the conditions under which a dynamic view of evolution might be supported (see Calcott and Sterelny 2011). Within wider contexts of both historical timescales and shifts in organizational hierarchies a diachronic pluralism of individuality is more than wild speculation.

However, if evolution is itself a stable process, then presumably the role of individuality remains static, which constrains our typology. It's reasonable to take a theoretical snapshot too. Clarke (2010, 2013), for example, identifies a functional role for individuality realized through multiple policing and demarcation mechanisms, and is also working within a particular evolutionary framework.

Clarke does venture into broader contexts in "Origins of Evolutionary Transitions" (2014, 314) working with multi-level selection models to track when "solitary organisms become subsumed within a new higher-level organism, which participates in a higher-level selection process." However, when discussing *origin*, rather than evolution, of individualities there exists a difficult metaphysical gap to bridge.[36]

The major transitions literature develops origin narratives not necessarily presupposing prior existences of particular individuals. That is, when there's

---

[35] Insofar as other types of nonevolutionary biological individuals are the products of evolution, then they too may be subject to change over time.

[36] Jim Griesemer shared this insight during a SPASHS meeting about origins and evolution of institutionality with his view of the three C's (coordination, cooperation, and collaboration) in cases where origin stories—whether of institutions or of individualities—don't presuppose the existence of them—how did the institution or individual originate?

interest in a particular form of individuality, it is often asked: how did that individuality originate? Analyses are conducted in contexts where expressions of individualities already exist, but the question remains of how that sort of organization began.

Concepts of conflict and cooperation are used to bridge both directions of the "one and many." Two origin stories are tracked:

(1) One making many (e.g., such as fraternal allegiance among daughter cells that stick together),

(2) Many coming together to form one (e.g., such as archaeon and protobacterium coming together to form a eukaryotic cell).

If an ontology is desired beginning with one, you'll look at things one way. If an ontology is desired beginning with many, you'll look at things another way. For most of the practical problems concerning transitions in individuality (e.g., in labs, field experiments, etc.) many expressions of individuality already exist and so origins (and the evolutions) of individualities become local empirical questions to solve.

3. Closing Remarks on Disciplinarity and Classification

In closing, there are some key questions of individuality pluralism that remain open for future investigation. How do we best understand a plurality of types at a time (i.e., a synchronic plurality) or over time (i.e., a diachronic plurality)? What sorts of local empirical cases can better inform accounts of origins and evolutions of biological individualities?

I maintain that many answers to such questions are shaped by, and conceptually contained within, the "edges" or boundaries of disciplinary domains and subspecialties. Others exist at the interfaces of those disciplines as they change over time. In all cases, there must be a robust and recalibrating pluralism with the capability to respect its own range and limits within and among its domains. That is what it means to conduct domain-driven analyses of the sort advocated for in this section.

Table 1 began this discussion. It represents an idealized ontic landscape because the chosen port of entry is somewhat arbitrary. If Table 1 was historically constructed, that would complicate the typology of subjects. Ditto for arranging primarily by discipline or profession. So, how are we to understand pluralism and biological individuality? Inspired by Wimsatt's (1972, 68) "Complexity and Organization" consider the following:

> Given the difficulty of relating this plurality of partial theories and models to one another, they tend to be analyzed in isolation ... but these incomplete theories and models have ... impoverished views of their objects ... It is as if the five

blind men of the legend not only perceived different aspects of the elephant, but, conscious of the tremendous difficulties of reconciling their views of the same object, decided to treat their views as if they were of different objects.

While an array of individualities was each analyzed in isolation to broach the topic, it is mistaken to approach biological individuality as merely an aggregate concept glossing over the complexities exemplified by the table's disciplinary entry points. As Wimsatt directs us (albeit on a different matter, but I take the same messaging to apply), complexity is in part located at the *interfaces* of those domains.

Those interfaces cannot be understood without considering plurality as a thicket in the Wimsattian sense: when navigating the weeds, it's about how you strike that machete. Any entry point to our landscape is frayed around the edges; disciplinary machetes do not have sharp blades. For example, Child was writing before the modern synthesis "hardened" when some disciplines were ignored or simply left behind. Thus, one can draw pre-synthesis insight from Child who drew from an array of specialties concerning metabolism, immune function, and growth occurring in developmental cycles. While analyses are domain-driven, one type of biological individuality is not exclusive to some subdisciplinary area.[37]

Biological individuality is held accountable to disciplinary ambiguity, changing disciplinary boundaries, and ways in which classificatory systems are affected by said ambiguities and frankly evolutions of disciplines themselves. And so, Table 1's taxonomy of individuality types was arranged according to domain; evolutionary, immunological, ecological, metabolic, developmental, and so on, which remains somewhat misleading. If domain-driven analyses of biological individuality are derived from the life sciences, the resulting pluralism must be considered against the thickets of disciplinary backgrounds and historical timeframes in which they are conceived.[38]

2. Critics & Methodology

Introduction to Section 2

> . . . we are responsible for boundaries; we are they.
> — Haraway (1991, 180)

Section 1's ontic landscape generated different types of biological individualities. That resulting pluralism was sorted and discussed within a broader context

---

[37] Kaiser and Trappes (2021) argue that biological individuality represents multiple, interconnected questions that altogether form a problem agenda.

[38] There are multiple routes of investigation in this Element. At this point, readers might consider moving to Section 3 for historical analysis of individuality before returning to Section 2's methodological focus or continue with the linear reading.

concerning the history of life. I called these approaches "domain-driven" because their analyses of biological individuality are derived from selected disciplinary domains or subspecialties of the life sciences.

Recall the thesis of this Element concerning biological individuality's value. One way philosophers tend to think about values across the sciences includes epistemic or knowledge-based values about theory, reasoning, success, and so on. Thus, Sections 1 and 2 develop around the theoretical and methodological aspects of biological individuality, including its role in the production of scientific knowledge. Critics of the work analyzed in Section 1 raise epistemic questions like what value, if any, does biological individuality have in the production of empirical knowledge?

Such critical approaches are "practice-based" because their attention directs to how biologists in lab and field contexts use and think about biological individuality in their work. For example, one might explore how individuating actions, like boundary identification, serve the aims of an intended experiment, how individuating activities define some term, establish an area for observation, or distinguish some observable phenomenon from others of a similar sort. The work on biological individuality discussed in Section 2 saturates a different conceptual space by drawing from epistemological, pragmatic, and methodological views of individuality's value in the scientific process.

I start by distinguishing three ways to take a practice-centered approach to biological individuality. Thereafter, the preoccupation with phenomenal qualities of biological objects, such as their boundary conditions, is critically analyzed. Prescriptive recommendations are then provided for newcomers to avoid remanufacturing standard puzzle cases against received concepts of biological individuality. To that end, Section 2 closes with an example concerning an underserved technological context before turning toward the social and political ideologies that have historically shaped the concept in Section 3. Biological individuality is not, and never has been, value-free.

Practical Critics from Epistemology

Section 2 takes a practice-based turn criticizing the general methodology of the previous section. Section 1's ontic landscape distinguished a plurality of biological individualities according to domains of the life sciences, but so what? What work can individuality concepts do for *us*? Let's briefly distinguish three different practice-based approaches to biological individuality.

Recently, there was criticism of individuality's relevance to biological practice (Kovaka, 2015; Waters 2018). Kovaka (2015) argued that the quality of empirical work on biological individuality is not determined by a resolution of the philosophical debate over individuality. Additionally, Waters (2018) called for a shift in focus. Rather than asking what a biological individual *is*, philosophers should ask how biologists conceive of individuals. These critiques sparked philosophical work tracking individuation techniques and processes in scientific practice. How do we approach biological individuality from a practice-based perspective?

There are at least three ways to approach biological individuality from a practice-centered perspective. First, consider an anthology that carries a certain thread of investigation to fruition and highlights how intuitive concepts of biological individuality, such as part–whole relationships, boundary delineation, identity, and continuity are in tension with practice (Bueno et al. 2018). For example, Kaiser (2018, 65) uses paradigmatic examples of parthood from biology to support her view of part-whole relations. In the same volume, Fagan (2018) explores the potential for consistent delineation and identification of stem cells in experiments. And Millstein (2018, 298) investigates the interdependence of land communities to make the case for the individuality status of land communities. Generally, approaches like Kaiser's, Fagan's, and Millstein's move from philosophical concepts to puzzle cases,[39] or alternatively, from the puzzles themselves to conceptualize about some feature of biological individuality and associated ideas like parthood, mereology, and identity. Therefore, one way to begin a practice-centered approach is to investigate how current individuality concepts (e.g., evolutionary, immunological, ecological, etc.) are used at different points in the scientific process. That investigation might concern how "common sense" and/or metaphysical criteria are (or are not) deployed, discovered, or satisfied in that process.

A second way to take a practice-centered approach is to study how philosophical debates over individuality matter for producing empirical products. However, what is at stake empirically was recently criticized by Kovaka (2015). She made a compelling case that while the question of biological individuality is relevant to the empirical study of biological processes, biologists do not need a resolution of that debate. Empirical work might include investigations into multispecies associations like bacterial biofilm communities or macrobe–microbe relationships (e.g., the human and microbiome), "super-organism" characteristics of eusocial insects, and so forth. To establish *the* criteria for what it takes to be an evolutionary individual – that is, the contemporary sense of objects of selective processes – has no empirical traction. Settling on the

---

[39] Section 1 explored an ontic landscape with regard to puzzle cases, now it's about making the epistemic moves more explicit.

necessary and sufficient criteria a priori does not matter for what biologists actually do or care about.[40]

Around the same time, and arguably in response, a slew of practice-oriented research followed that discusses both the historical and experimental roles of individuals in scientific reasoning (Lidgard and Nyhart 2017; Bueno et al. 2018). Philosophers are now well beyond advocating for a singular definitional context concerning biological individuality: stereotypical organismal traits are rejected that constrained philosophical criteria of the past,[41] and fields of biology are approached with various points of focus. However, I take there to be a shared aim: understanding to what extent biological individuality – including established philosophical concepts, traditional ideas about individuality, identity, and related metaphysical ideas, and potential criteria for individuating activities – matter for products of empirical work and whether it really tells us about how nature is organized.

Though asking why individuality matters and why it's important for the work of biologists was an angle previously taken by Clarke (2010, 2013). Clarke argued that individuality matters for biologists insofar as they need to count. That counting activity requires the ability to distinguish offspring consistently across a set. She argued that individuality matters for the population biologist's need to determine fitness of a population. In fact, Clarke *frequently* motivated her papers by framing how individuality matters for products of empirical work. Waters (2018), while critical of Clarke's overall metaphysical approach, analyzes the conditions under which individuation practices matter for geneticists when delineating genes. In both cases though, and perhaps in many others, why philosophers should care about individuality is motivated by if and how it matters for the empirical *products* of some area of biology – the outcomes of scientific inquiry. While Kovaka (2015) criticized certain approaches to establish that connection, it is precisely that kind of philosophical research programme I would like to draw attention to. In other words, one research program for philosophers of biology is to determine whether and how individuality – with its philosophically-charged history – functions in producing empirical results. I take this as roughly an effort toward identifying individuality's place (or lack thereof) in the structure of scientific reasoning about targets of study.

On the one hand, the connection between individuality and empirical products can be addressed by investigating the potential need for individuality

---

[40] How is the empirical traction of philosophical debates measured? By citation impact? If so, Ereshefsky and Pedroso's 2015 paper has seventy-one citations on Google Scholar, many (if not most) from biology and health science.

[41] However, some rebut that ignoring distinctly organismal traits, like agency, is to our detriment. See Walsh (2018).

concepts (and clarifying them). On the other hand, we can also explore how individuation techniques help biologists do their research, and specifically which techniques help to identify the phenomenal qualities of objects in their domain. The direction of inference here is very local and specific to the field or subfield of biology and the empirical work for that field. And so, one role for philosophers of biology is to investigate the connection(s) between individuality concepts and/or individuation techniques and the empirical results that follow. That is our third way.

In summary, I have covered at least three different practice-centered philosophical approaches to biological individuality:

(1) Deploying philosophical concepts to understand puzzles from biology.
(2) Using puzzle cases from biology to test traditional or a priori philosophical assumptions about individuality (e.g., monism and conceptual analysis via necessary and sufficient conditions) and using puzzle cases to generate new philosophical views.
(3) Exploring what role individuality concepts and individuation techniques have in the production of empirical results.

The list above is not exhaustive as it provides merely a preliminary inventory. While (1) and (2) demonstrate a particular relationship between philosophy and science characterized by the direction of reasoning (i.e., from philosophy to the science or from the science to philosophy), (3) digs into the process of scientific activity leading to empirical results as products of that process.

Phenomenal Qualities, Objects, & the So-Called Practice Turn

While Section 1 was theoretically motivated by metaphysical concerns about life's history, a turn to epistemology of individuality carries pillars left from early modernists, positivists, and their aftermath. McConwell (2020) suggests that hallmarks of positivist philosophy of science haunt the "practice-turn." Work on individuality is revealing in that regard, for example, by reconstructions of how intuitive properties from metaphysics fit into the structure of science, and how empirical cases challenge expectations. These claims call for some unpacking and justification. What exactly are those pillars of the past?

1. Pillars of the Past: Colonial Logic & Objectification

Sinclair (2020, 67) discusses how biological individuals are taken for granted on the ground level of many (philosophies of) sciences. There is critical overlap between focus on individuality and problematic enlightenment metaphors, concepts, and values. The tendency to "see the world as composed of essential, clearly delineated individuals is part of a wider settler-colonial philosophical project that justifies the subjugation and attempted erasure of Indigenous relational science and knowledge" (2020, 67). By referencing the enlightenment, Sinclair notes a pillar of the past, but what's the problem?[42] Consider the following to help illustrate the point.

Sinclair discusses how indigenous logics emphasize relationships and non-binary natures. That is in stark contrast to asking the following: Are puzzle cases for individuality, such as ecological communities, individuals *or* not? A binary logic frames the expected answer as only one disjunct of an exclusive "or". That binary expectation has social and political consequences, especially in conservation efforts that assume only singular, unified organizations or "individual agencies" can be objects of recognition and protection. This is part of why I take Aldo Leopold to have sought an organismality (or as Millstein has argued an "individuality") designation for land communities. In that sense, individuality as a designation translates to recognizing those rights and protection.

In other words, colonial logic has *normative* consequences for ontological designations of individuality. For example, Millstein's work on Leopold's land ethic suggests that the individuality status of land communities (i.e., both biotic and abiotic components) is a prerequisite for moral consideration in conservation efforts. Individualism as an organizing principle in colonial frameworks protects ecological "agencies" by reducing restraints on freedom and by recognizing eco-interactions and natural systems as entitled to special privileges. That is, if forced to work within the binary logic of a colonial past, then designating land communities as individuals will secure both freedom from external disturbances that impede functioning (e.g., like polluting, commercial use, etc.) and protection in the form of persistent security. As one singular, unified individual (i.e., the land community in this case), it has needs capable of recognition, such as to persist on its own terms without disruption, and protection of those needs.

Similarly, the status of individuals as objects of recognition and protection (or conversely, of *denial* and *extermination*) applies to cases of conservation through management of the environment. In colonial frameworks, that management is often viewed as a one-way relationship of control, rather than a reciprocal and embedded process of humans alongside and within nature.

---

[42] The relationship between the enlightenment and origins of modern colonialism requires discussion that cannot be provided here. A referee aptly points out that those who historically endorse or do not endorse binaries in their philosophizing does not map onto colonial powers and indigenous groups, respectively. Rather, there are feminist and post-modern deconstructions of binaries, which do not necessarily derive from indigenous knowledge and scholarship and vice versa.

The former separates the manager (i.e., humans) as external entities imposing their will often for use and exploitation of the land and by that action objectifies nature. This suggests that determining the individuality status of some ecosystem, ecocommunity, or Leopold's land community, and so on for conservation and protection efforts is itself a historical product of nature's colonial objectification. That objectification is driven by binary assumptions about how individuality as an organizing principle is applied precisely for (human) management and control of nature.

2. Individuating Characters & Their Phenomenal Qualities

Undoubtably, individuality is a scientific tool to *objectify* the natural world: what are biological individuals if not *objects* sorted and typed by systems of classification. Those systems organize where distinguishable organic units belong under those schemas. Promoting epistemic traction on nature's complexity are qualities that make organic objects readily identifiable. For example, consider the slender body shape of hornets and wasps compared to bees, the varying texture of mountain pine needles compared to red pines, the red marks as "ears" on red-eared slider turtles as a distinguishing feature compared to painted turtles, and so on. Such phenomenal qualities – individuating characteristics from the physical and behavioral markers of bone shapes to the pitch of howls signaling aggression – help naturalists identify, track, and sort according to the systems of organization existing in whatever field of biology.

Philosophers already discussed individuating characters and the discerning work they do in phylogenetics, and the nature of characters as basic units of biological analysis. However, the epistemic value of individuating characters is not only how they historically trace degrees of relatedness in descent, but also how they provide material for certain adaptive functions (Griffiths 2006). It is "confusing that 'character' can mean either any measurable property of an organism or only a property recognized as biologically significant in some theory of the organism" (Griffiths 2006, 12). The ambiguity of 'character' becomes more complex when not just of single-species organisms, but also of multi-species collectives spanning organizational scales. However, Griffiths clarifies that philosophers are familiar with the same ambiguity in 'property'. Whether measurable or theoretically significant or both, 'character' is used in a highly specialized sense – they are the phenomenal qualities put to use, measured, and manipulated. In that sense, they are tracking tools for the ways in which individuals are *phenomenally objectified*, rather than considered active subjects.

In contrast to prior work on character individuation, phenomenal qualities associated with biological individuality shift slightly. Phenomenal qualities for discerning biological individuals range from part–whole relationships to boundaries based on both material borders and/or functional limits.

For example, while a mitochondria's functional capacity can be limited by the bounds of its cellular container (e.g., its mobile capacity limited by that material boundary to move between cells under certain contexts), the functional interactions, rather than material boundaries per se, of keystone species with other members of an ecological community identify the bounds of a functional group.[43]

Other phenomenal qualities, such as those detected through immediate experience or with the help of a microscope or other tool, are tougher to get a handle on, and so philosophers ask whether the one-to-many symbiotic relationship between a Bobtail squid and their illuminating bacteria is consistent enough to consider that consortium as one, continuous individual. Or, concerning matters of identity, they ask whether the continuous turnover of our cellular parts challenges both the oneness and sameness of who we are.

And so, the phenomenal qualities just discussed – *part–whole relationships, boundary closure (both material and functional), stability, continuity, and identity* – can be contrasted against how character individuation decomposes objects. The listed qualities are distinctly philosophical. As analytic categories – the lenses of analysis – they harken back to the hallmarks of concepts like identity and discernibility that shape philosophy's past as a metaphysical enterprise. Past pillars indeed.

Philosophy, even in the practice turn, runs biological objects through the same ringer, through the same categories of analysis Sinclair (2020) points to when emphasizing the enlightenment's long shadow. I am critical of the assumption that determinate properties of individuals are (and will) manifest in their concreteness: that an individual's individuate-*ability* is something rather distinct from how they are conceived. Pre-theoretical concepts of phenomenal qualities do not always harmonize with the shifting, processual nature of biology, nor in how those qualities function in a binary colonial logic. So, individuate-abilities must be operationally fluid.

If rejecting *positivist* standards (e.g., standards of necessary and sufficient conditions, uncovering universally absolute truths from scientific reasoning and data), then exactly what standards govern conceptual analyses of biological individuality? Products of our analytic categories are at times familiar, and sometimes surprising. Bueno et al. (2018) argue that assumptions shaping categories of analysis like mereology and boundary closure are challenged by

---

[43] Presumably though, if functional interactions of species are constrained by biogeography and geological factors, then material borders still matter.

the way biological practice is conducted. But one consequence is that a puzzle's status *as a puzzle* of individuality is in part determined by our analytic categories and in part determined by the way scientists conduct empirical work.

In conclusion, phenomenal outputs are characterization-laden: they refer to abstract ideas, and are learned, taught, and housed in the kind of disciplinary content that many would like to see deconstructed and even dissolved in philosophical analysis, such as colonial and positivist frameworks. Certainly, disciplinary practices change. Yet isn't one promise implicit in the so-called "practice turn" that we're turning away from such pillars of the past? If so, there may be reason to receive that promise with some caution.

Turn-talk punctuates history into before the turn and after: it divides practices, mechanisms, processes, concepts, and theories into those two select categories, for example, practice-based vs. not (see Griesemer 1996a, 1996b). The risk is, of course, homogenizing stages as if some kind of categorical shift occurred. But history never really stacks so neatly. "Turn-talk" is *branding*.[44] It marks resistances to what came before. But it must be asked, what exactly is resisted and where to next? What pillars of the past should be left behind when analyzing biological individuality?

Analytic categories concerning absolute natures, boundaries, and mereology in philosophy are infused with past folly and foible.[45] If those categories continue to shape the way individuation activities across the sciences are reconstructed and epistemically evaluated, then the "turn" will continue its circular direction around the conceptual axis Sinclair is worried about. Epistemology of biological individuality is not conducted in a history-free vacuum.

A Cottage Industry

I just diagnosed the "turn" toward practice-based approaches to biological individuality as still carrying pillars of modernism – a complex of enlightenment, colonial, and positivist ideals. Next, I turn to prescriptive discussion: What newcomers to the topic of biological individuality should avoid, and an example of a socially relevant and underserved area of biotechnology.

---

[44] Jim Griesemer ties the concept of branding to "practice turn" talk. Arguably, Kuhn used turn-talk against the logical empiricists, the logical empiricists made a similar claim against the logical positivists, and the logical positivists, as Klein argues in "philosophy at war: nationalism and logical analysis" (2020), made yet a similar claim to recapture meaning that was hijacked by nefarious regimes.

[45] For an example, refer to Section 3's discussion of biological individuality's dark side; how the concept has been shaped by social and political ideologies about progress and perfection. Biological is not, and has never been, value-free.

Metaphorically, avoiding a cottage industry and reworking tactics calls for reworking blueprints and plans. While there's been much work to establish a plurality of individuality types – immunological, ecological, different versions of evolutionary, and so on – applying that type's criteria to puzzle cases is now standard fare. For newcomers, the pedagogical value of these cases is undeniable to test the scope and limits of our conceptual criteria. However, a cottage industry should be avoided where puzzles are consistently remanufactured in our own disciplinary home. Compound organisms and associative relationships show how nature often confounds pre-theoretical (and perhaps even theoretical) expectations. However, as philosophers become interested in the activities of scientists and their communities, one can reasonably ask where to go from here. Some different tactical moves are needed.

Using nature's puzzles to test concepts is not an activity conducted only by philosophers though. Scientists do it too. Scott Gilbert's et al. (2012) "We Are All Lichens" paper is titled with an intuition-bashing slogan and works toward exactly that; intuitive individuality concepts are worked over puzzle cases. To be clear, that paper serves as an example of interdisciplinary theorizing about associative relationships among organisms. However, I do not take that paper to be an instance of a biologist working over those concepts *in the field or lab* per se. Although, scientific practice does include more than the application of technical skills, instruments, and sets of controls. Conceptualizing and theorizing about the big ideas is part of it too. In this case, Gilbert et al. aptly draw on the phenomenal disconnect between our pre-conceived notions of individuality and what is found in nature. It is counterproductive to be overly hard lined about what exists within the scope of scientific activity; both theoretical and empirical activities count.

From a philosophical standpoint, however, situating our own analyses of biological individuality within the context of practicing biologists calls for a different tactic than using published products to test conceptual analysis. Another way to ask that question is: What is the epistemic relationship between conceptual analysis of biological individuality and scientific practice?

To argue that biologists need a clear concept of biological individuality for empirically successful work assumes that biologists move from a top-down approach where a concept is adopted *a priori* and applied to some association or system. While Kovaka (2015) argued that empirical work still gets done regardless of conceptual agreement, Kranke (2021) builds on that account and utilizes concepts as *models*. Her model-centric approach combats any deep metaphysical line one might assume from regular old conceptual analysis (i.e., such as necessary and sufficient conditions for some individuality type). Concepts-as-models do not carve nature at its joints, but develop within certain

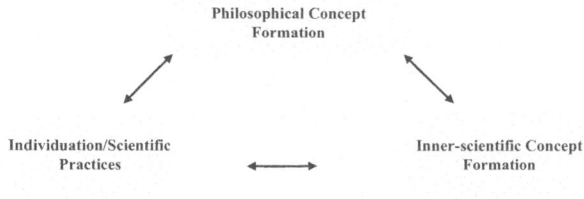

A revised depiction of Kranke's 2021 depiction of a
model-centric approach to conceptual analysis

Figure 7 Kranke's original depiction titles the top as "Meta-scientific concept formation." I shifted the description to "philosophical concept formation." Overall, a model-centric approach keeps concepts as living, rather than static, entities that adjust as scientific practices refine and change

limits; working parameters that can change as scientific practices recalibrate working concepts as depicted in Figure 7.

So, conceptual analysis can be revised to better capture how biologists identify the boundaries of biological systems (e.g., such as Kranke's test case of host-parasite systems). In effect, a model-centric conceptual approach is a different tactical move and one that is mindful of pluralist sensibilities. Kranke's survey of studies followed by adjustable analyses proves useful as a different way to work puzzle cases according to concepts. Regardless, I do find myself wanting something more than a model-centric approach to conceptual analysis, while at the same time acknowledging how it can reveal disconnects between pre-theoretical and theoretical individuality concepts and their (partial) use (or disuse) across science studies.

A second assumption that Kranke addresses runs a bit closer to what I have in mind for a tactical retooling. Some might view science studies as aiming to (re) construct useful concepts for scientists with the potential to guide and evaluate their empirical work. Kranke proposes a pragmatic dimension. Certain concepts-as-models (i.e., concepts that are workable under certain conditions and capture practice in a piecemeal way) emphasize different epistemic aims and purposes in science. However, that is exactly the prescriptive tone of conceptual analysis conducted in philosophy that is of interest, and perhaps what was targeted by Kovaka. That is, individuality concepts are often characterized with a normative explanatory role for science. This is where things become woefully complicated. Let's focus in.

On the one hand, philosophers might want to learn *from* scientific investigations of biological individuals in the world, rather than only focus on what scientists can learn from stabilized conceptions of individuality. Other disciplines might be consulted for guidance on how to structure philosophical observations of scientific practice to generate conceptual analysis from the ground-up.[46] Specifically, socio-philosophical work on biological individuality is to immerse oneself *as a philosopher* in lab groups and fieldwork to better understand how individuating practices succeed and fail in local contexts.

Presumably one hope for the above strategy would be to produce wider conceptual applicability in the sense that Havstad (2020) grappled with concerning generalization from ethnographic lab studies. The epistemic value of quasi-sociological/STS approaches with philosophical edge calls for structuring one's role as a participant observer: philosophers sometimes share similar theoretical questions with the scientists (in this case biologists) whose practices they wish to study. As the "practice turn" in philosophy of science deepens, upcoming scholars should consider the ways in which they can intentionally shape their interactions with scientists; go beyond the published products and dive directly into the community action, while reinforcing that entry point *as philosophers* in those settings.

On the other hand, and more substantively for present purposes, prescriptive and evaluative conceptual practices may be controversial when applied to some empirical work, however, normative input is in high demand depending on the context. Plainly put, more underserved contexts must be identified. And biotechnology is a space where questions of individuality, identity, and values coalescence into a rich and underserved area insofar as the topic of biological individuality is concerned.

For example, in literature on kinds and classification, a parallel intellectual arena, Kendig and Bartley (2019) focus on kind-making in synthetic biology – a multidisciplinary area of work that concerns the redesign, engineering, and manufacturing of biological systems and parts. They explore how this area of research, which includes both academic and industry interests, impacts philosophical conceptions of biological parts and kinds. A synthetic (versus native) kind is defined as a form of life (or life-like thing), which is constructed by human-assisted engineering. For example, "an *E. coli* population harboring a synthetic DNA plasmid might be considered to be a synthetic kind"; a synthesis between the wild-type organism and the synthetic plasmid (2019, 81–82). Besides kindhood, it remains to be seen how synthetic biology might inform not only biological individuality and how it functions in scientific knowledge, but also what it *could and should be* from a design standpoint.

---

[46] Exploratory scoping as a means of philosophical investigation can be inspired by grounded (i.e., starting from the ground-up) theory in sociology (e.g., Gerson 2017) to structure philosophical data from time spent in labs and fields.

While individuality, biotechnology, and ethics serve as a topic for the next entry, something more general can be said about the two possible directions just mentioned.

To be clear, I suggested two ways to avoid a cottage industry of work on biological individuality:

(1) Immerse oneself in structured engagement with scientific processes beyond published products (i.e., go hang out with biologists as a philosopher and structure how you generate/analyze philosophical data from those interactions),

(2) Consider normative questions about individuality in new and underserved spaces, like biotechnology.

These two approaches go beyond preoccupations with boundary conditions of phenomenal objects – the subject matter of the previous critique. Instead, *analysis should be structured* through intentional interdisciplinary contact with local scientific communities. Think beyond the boundaries (in this case, disciplinary boundaries).

The previous essay critiqued how pillars of modernism haunt the "practice-turn" concerning biological individuality. What does that mean for present purposes? Those pillars include necessary and sufficient boundary criteria, specifically whether those criteria map reality when applied to systems, associations, and units broadly speaking. Instead, a different approach should be taken: Consider *individuation* in an interdisciplinary context: Why does individuality matter for scientists and their research communities? How can individuation practices inform policies for research values and integrity? Some might be concerned that socio-philosophical work borders on science reporting. I counter that interdisciplinarity does not devolve philosophy, but enriches its milieu. That is the real promise of the practice-turn insofar as biological individuality is concerned.

Next, an example of a previously underserved and socially relevant context is considered before the final Section 3 of this Element.

A Space for Technology & Values (Coda)

As previously mentioned, biotechnology is a space where questions of individuality, identity, and values coalescence into a rich and philosophically underserved area. That statement needs some context. It supports both a backward and forward-facing sentiment toward the intersection of philosophy, biotechnology, and values. Let me explain by first looking at where we've been.

1. Looking Back

Those specializing in science and technology studies, including philosophers, have already drawn conclusions about individuation and identity based on technologies produced from applied sciences to navigate and observe nature, while also trouble shooting medical needs. As Donna Haraway states in *Cyborg Manifesto* (1991, 163, 180), intervention efforts focus on control strategies concerning interfaces:

> One should expect control strategies to concentrate on boundary conditions and interfaces, on rates of flow across boundaries – and not the integrity of natural objects . . . no objects, spaces, or bodies are sacred in themselves; any component can be interfaced with any other if the proper standard, the proper code can be constructed . . . we are responsible for boundaries; we are they.

Haraway tells us that the agent has an active role in organizing boundaries of (or in?) their worlds. Taking seriously the deconstructive element in Haraway's work – pulling apart preoccupations with stable and sharp boundaries – then agents are responsible for where those boundaries rest at any local point of time. The heavy philosophical lesson, as I take it, concerns the relational status of boundaries inspired by both techno science and science fiction: foundational distinctions are contingent at best. That is due to how biology and politics are intertwined, Haraway's *biopolitics,* through feminist reflections on social stratifications (i.e., race, gender, sexuality, class, etc.) in science and biotechnology. A main lesson from Section 1 of this Element concerned pluralities of individualities as domain-driven, that is, as defined according to some disciplinary area of the life sciences and held accountable over time to changing and ambiguous disciplinary boundaries. My intention in returning to Haraway is to emphasize how practitioners have an active role in organizing material and ideological aspects of those boundaries. Consider the following:

> . . . biotechnologies are the crucial tools for recrafting our bodies. These tools embody and enforce new social relations . . . technologies and scientific discourses can be partially understood as formalizations, i.e., as frozen moments, of the fluid social interactions constituting them (2016, 33).

In other words, the agent is an active contributor to boundaries imposed by technology. Additionally, though, as emphasized in *Companion Species Manifesto,* that role is entrenched in the history of companionship with other species – be it dogs or microbes. Our participation in these relationships is shaped by both nature and socio-cultural ideology.[47]

---

[47] One might consider a similar lesson from Latour and Woolgar's work in the 1970s studies of the Salk Institute that criticized the presupposed distinction between the natural and social. Resituated in this discussion, synthetic individuality includes material *and social* factors that

Haraway draws conclusions that undermine characterizations of biological individuals apart from their social contexts. And that is exactly what piques public interest now: *considering individuality at the complex juncture of bio-social spaces.* There are many popular science articles that demonstrate this channel of interest begging for technical philosophical input. Here are three examples.

Wallace-Wells (2015) published "Adventures in the science of the super-organism" using examples to motivate why we should care about biological individuality, such as one twin ingesting the embryo of the other twin in utero, the trillions of gut bacteria that house themselves in/on us, the viruses that colonize our DNA, and the ever-so-strange fetus-in-fetu cases. What is at stake, according to the author of this public-facing op-ed, is reconciling our own personal individuality with the notion that humans are super-organisms with mosaic identities. However, Wallace-Wells is concerned about a richer sense of individuality – its social sense – as if that hinges on biological boundaries. This line of bio-social reasoning is taken even further in a different article published the same year.

Salisbury (2015) in *Science Daily* asks whether the pronoun "I" is obsolete due to recent microbiological research. The proposal is that holobionts and hologenomes – associations between macro and microorganisms are the units of biological organization subject to selection, drift, and mutation. And while there has been philosophical work examining various forms of scientific pluralism using holobionts as a case study (Şencan 2019), Salisbury presses the social consequences of a mosaic self. Searches on *Aeon* and *Quanta Magazine* will produce similar results concerning the bio-social interest that shapes the way individuality translates to both cross-disciplinary and public realms. Enter technology into the equation and the complexity increases though.

Donahue (2017) published a *National Geographic* article titled "How a color-blind artist became the world's first cyborg." The article covers the case of Neil Harbisson, a color-blind artist, who uses an antenna-like implant that allows him to "hear" color via vibrations in his skull. While certainly, as the article states, humans have used technology to alter physical and mental capabilities, the claim of the article is that we're already cyborgs in Haraway's sense. Harbisson views the antenna as just another part of his body: we alter both ourselves and our environments. So, when Haraway says we are responsible for boundaries – that we *are* the boundaries – this science fiction reality is what I take Haraway to have in mind.

---

cannot be strictly divorced from natural phenomena. As Havstad (2020, 23) discusses, the (social) construction of scientific facts in the lab remains one of the most controversial aspects of those studies. Applied to the present case, there is a dependency relationship between synthetic individuals and their respective fields of scientific practice.

Popular op-ed articles are watery signatures of philosophical concepts suspended in time: they take bits of philosophy that look interesting from the outside and press on questions that sometimes philosophers no longer want to touch. In other words, one might object that popular science articles about social questions concerning individuality in biotechnology may not be motivating for philosophers. To that end, Kendig and Bartley's (2019) work on synthetic kinds could have been further developed as an analogous way to approach synthetic individuality. Or controversies surrounding human gene editing and the birth of the first three parent human individual in 2016 could have been considered.[48]

In response, I maintain that how popular science articles frame social questions about biopolitics and biotechnology should be investigated, especially if philosophers of science have a responsibility to contribute to public understanding of science beyond what science journalism provides (see Cartieri and Potochnik 2014). Philosophy's social relevance includes not just educational and institutional reform, but also management of science's social position, which involves its public expressions by both journalists and scientists.[49] Popular biotechnology articles are one main avenue through which publics will access questions about biological individuality. Thus, social questions drawn from those sorts of articles exemplify how science fiction, synthetic biology, and politics merge *for the publics* in bio-social engineering of individuality.

While the relationship between biology and society is informed by knowledge systems that define those spheres (i.e., biology and society, science journalism, fiction, and public education), that juncture is approached by philosophers of science and biology in other areas, such as evolution's meaning and science's sociality. Biological individuality is just another space in which public-facing work can be fruitfully repurposed beyond traditional questions of metaphysics and epistemology. Looking forward there is much opportunity for a "science and society" approach to biological individuality.

2. Looking Forward

The normative consequences of individuality and identity are significant in contexts of biotechnology from organoids and chimeras to CRISPR-Cas9

---

[48] See Zhang and Wang (2016) concerning gene editing of human embryos in assisted reproductive medicine. The specific case referred to involves removal of DNA from the nucleus of ova affected with a mitochondrial disorder. That DNA was inserted into a donor's ovum without the problematic mitochondrial DNA and fertilized in vitro before being transferred to the uterus.

[49] Another example of why popular expression matters: JBS Haldane in *Causes of Evolution* (1932, 165) cites Stapledon's dystopian novel *Last and First Man*. Beyond technical scientific work, Haldane was a prolific popular writer and proponent of eugenics. That science fiction influence on the history of the modern synthesis has been under-explored.

gene therapy in biomedical research contexts.[50] Organoids are biological structures created from stems cells *in vitro,* which partially mimic the function of organs, while *cerebral* organoids are "stem-derived 3D biological structures, which are self-organizing in morphological units resembling a developing brain" (Hostiuc et al. 2019, 119). How do these cases fit into systems of values? Cerebral organoids are of special interest because they derive from human embryonic or adult cells but are subject to the use of techniques that can potentially alter human constitution to include genetic material from other species resulting in chimeras. As Hostiuc et al. (2019) state, human-animal chimeric organoids have already been developed and chimeric cerebral organoids could be a next step. They cite Karpowicz et al. who argue that these neural chimeras should,

(1) Use a minimum number of human stem cells possible, and
(2) Make sure that the host animal is not too morphologically or functionally similar to humans to mitigate the risk of developing human-like neurological networks (2019, 120).

These criteria are followed by attempts to define some threshold of considerability unique to humans to reduce the risk of causing harm, or at least to mitigate the risk of harm in unknown contexts (120ff).

Ethicists will no doubt groan at the threshold-unique-to-humans line and even the most scientifically inclined metaphysicians will recognize the tale as old as time: identity and (re)constitution. Replace the Ship of Theseus with gene editing technologies and Petri dishes and a philosopher will argue that we're still in that same boat. What's missing in Theseus's old ship parable, however, is *who* replaces the planks: as philosophers we tend to focus on the end result, the published planks, without paying much attention to who is doing the constructive work. So perhaps Theseus's ship needs an updated framework, one which aligns with Haraway's insights; namely, about *who* is reconstructing and deconstructing the boundaries that shape biological identity, alongside their motivations and methods.

Such a focus on scientific work – that is, on *who* individuates and how those individuation practices have philosophical consequences for the products of scientific knowledge – is not enough to inform the normative concerns cited above. More is needed from the science and values and STS networks to

---

[50] Normative interpretations of what synthetic individuals should be like or how we should (or should not) use them will depend on social narratives about the meaning of their "syntheticity." The prioritized value of human cells and tissues relative to non-human animal products discussed in Hostiuc et al.'s 2019 empirical work is itself an ideologically charged narrative about the priority of human value, which frames the (bio)ethics of synthetic individuals.

develop individuality's epistemic dimensions in conjunction with its social and political features across biotech contexts. Approaching a similar theme, Longino (2002, 59) has described the lesson I'm now trying to capture: " . . . nature does not, cannot, act alone in the laboratory . . . but needs a whole social and institutional structure in order to be deployed." The point stands in both experimental and bioengineering contexts: synthetic individuality draws attention because of the immediacy of human intervention, but it is not necessarily unique in either (1) its dependency on disciplines or domains as was discussed in Section 1 of this Element or (2) its relationship with social and political values, for example, like assumptions of human "specialness" as will be further discussed in Section 3's historical setting.

With the ability to shift the constitution of a biological entity's individuality through technology, there is promise for productive work ahead. Philosophers of science engage biotechnology and biopolitics in a way that both departs from, yet complements, the current work of bioethicists. Philosophy of science, even that which focuses on values and public policy, varies in nuanced ways from that of analytic and applied ethicist interventions. Looking forward, philosophical possibilities emerge from synthetic biological individuality, that is, not what biological individuality is or how it's used, but what it *could* be and how it *should* be designed for society.

3. In Historical Context

Introduction to Section 3

> The main fact abides – that progress is an evolutionary reality, and that an analysis of the modes of biological progress may often help us in our quest for human progress . . . The next great problem . . .
> —Huxley (1923, 90).

Julian Huxley (1887–1975), a prominent evolutionary naturalist, argued that accounting for biological progress is key to unlocking the direction of human evolution. Progress in evolution is a contentious concept, which I contend is intimately tethered to biological individuality in the history of evolutionary thought. Specifically, biological individuality has a dark side, which serves as a cautionary tale wrapped in social and political ideologies about progress and perfection in nature. This Element closes by defending the following claim: biological individuality has been used to promote politics of social ideologies about managing human evolution through use of the life sciences.

Up to this point, Sections 1 and 2 focused on theoretical and methodological aspects of individuality's value for gaining knowledge about nature. The leading question of this Element was a critical challenge concerning whether biological individuality matters in the production of scientific knowledge: Why does individuality matter for biology? For philosophy? *What is its value?*

Recall that philosophers consider values in science in two different senses. One concerns knowledge-based values about theory, reasoning, success, and how scientific knowledge is generated. There are two sets of lessons to be drawn in that regard:

Lessons concerning theoretical reasoning

- The Species-As-Individuals Thesis challenged essentialism about species taxa. Evolutionary theory demands species taxa be historical entities, rather than ahistorical natural kinds.
- Many types of biological individualities are generated according to domains and subspecialties of the life sciences, for example, evolutionary, ecological, immunological, metabolic, and so on.
- Individuality fruitfully facilitated debate over reproduction as a mechanism of inheritance considered materially and whether formal interpretations of transmission are enough.
- Individuality in immunology, ecology, and metabolomics generates discussion over matters of life, health, and disease (e.g., the role of the holobiome in autoimmune disease, and normative understandings of ecosystem health).
- There is not only a domain-driven plurality of biological individualities but also a plurality of types within types, such as different types of evolutionary individualities (a synchronic pluralism).
- Considering individuality against the history of life illustrates the production of new types of individualities over time (a diachronic pluralism).

Lessons concerning method

- Different ways to investigate biological individuality in scientific practice occur by direction of reasoning (i.e., from philosophy to the science or from science to philosophy) and by activity producing empirical products.
- The practice "turn" is pre-occupied by boundary conditions of phenomenal objects – their individuate-*ability* – risks unanalyzed assumptions from pillars of modernism (i.e., a complex of enlightenment, colonial, and positivist ideals).
- To avoid a cottage industry of work (i.e., using puzzle cases to test intuitions and conversely using puzzles cases to generate new types) structured interdisciplinary engagement with scientific communities is needed.
- Normative questions exist in new, underserved intellectual spaces, such as synthetic individuality in biotechnology.

And so, biological individuality is theoretically and methodologically valuable in the production (and the philosophy) of scientific knowledge. However, there is a second sense of value – "non-epistemic" value – concerning the social and political consequences of individuality. The rest of this Element is dedicated to that under-explored sense by drawing from naturalists like Darwin, Julian Huxley, Thomas Huxley, and Asa Gray.

Specifically, I show how biological individuality is a concept historically shaped by political and social ideologies with not only theological features but also worrisome eugenics overtones about biological individuality as a tool to control humanity's evolutionary future. Next, I develop a historical take on biological individuality's connection to assumptions about agency, design, perfection, and progress in nature. It is in that sense that individuality is not, and never has been, value-free.

Section 3 begins with a comparative analysis of individuality and agency in both contemporary and historical senses. How individuality was a driver of evolutionary progress is demonstrated by drawing from earlier conceptions of compound animals and zooids. As we'll see, together agency and increasing complexity shape Julian Huxley's view of individuality as a placeholder for progress: for Huxley *biological individuality drove progress*. Finally, the social and political consequences of that view are unpacked before some closing remarks.

Coloring Outside of the Lines

There is reason to reject the prospect of individuals as merely cogs in the nature machine; as passive objects with respect to processes and classificatory practices. Agential thinking treats biological individuals as active in constructing their environments and determining their evolutionary futures. Walsh (2015, 2018) takes a strong stance against passivity arguing that agency is a unique organismal capacity distinguishing organisms from nonliving things.

Historically, biological individuality's agency was not just about active natures of organisms though. It was accompanied by narratives about social progress in evolution writ large. The latter concerns our cautionary tale. To draw out that contrast, let's begin with an analysis of recent work resisting a social (i.e., intentional, conscious, or "inner" striving) characterization of organismal agency.

1. Two Contemporary Takes on Agency in Biology's History

At first glance, approaching biological individuality through an agency lens risks coloring outside the lines of biology. Connecting the biological to the social is controversial and treacherous ground.[51] For individuality, that means connecting basic boundaries of biological objects – whether distinguished by us or discovered "out there" – to social concepts like self-identity, intention, and meaning. Socializing the agency of biological individuality risks a vestige of biology's past. So, how might agential dimensions be considered without venturing too far down that kind of path?

Around the time that Walsh's 2015 book *Organisms, Agency, and Evolution* was published, it seemed in stark contrast to current debates over evolutionary individuality discussed in Section 1. The latter was an alternative research context focused on how inheritance occurs for individuals in selection, which did not include organisms as active participants in evolution per se. Walsh resists notions like intentionality and consciousness, especially in a recently coauthored exploration of the agency perspective (see Sultan et al. 2021). Working at the edge of the social, while at the same time resisting that dimension, marks an intriguing foray into organismality from an agency point-of-view.

Walsh's (2015, 2018) view embraces a traditional dimension of organisms; an active dimension that became less popular after the modern synthetic theory of the mid-twentieth century.[52] A return to active organismality marks an *extended* evolutionary point of view, but what exactly is the relationship between active organismality and an extended evolutionary view?

The processes of development are not just outcomes of some genetic script, but an outcome of processes happening within the organism on all sorts of (environmental and other) information (Sultan 2019). In discussion, evolutionary ecologist Sonia Sultan explained that causes of phenotypic variation must be distinguished from their record as DNA sequences. Many evolutionary experiments are designed to surpass environmental components, which naturally inflates genetic components (see Sultan et al. 2021). And so, on an extended evolutionary view the organism matters as the *locus of interaction* dependent on environmental contexts.

Sultan et al. (2021) illustrate how an agency perspective draws from a different conceptual framework. Mechanistic explanations address components of systems to explain how they work (e.g., the genetics), which is successful if a system works the same to produce the same outcomes across contexts. However, if that

---

[51] Controversial because of assumptions about biological determinism. See debates over the publication of E. O. Wilson's *Sociobiology* (1975).

[52] The modern synthesis combined Darwinian theory of evolution by selection with Mendelian inheritance in genetics, but also coordinated sciences relevant to evolution, which includes genetics, systematics, paleontology, and so on.

system is context-dependent, then a mechanism-based explanation will be incomplete producing gaps in understanding.[53] An agency perspective is not in competition, but rather complementary to mechanism-centered views: organisms integrate both genetic and environmental influences to produce a specific phenotype, and so how a phenotype works (or doesn't) for the organism is what needs to be explained.

This research program encourages the redesign of experiments to reveal complex evolutionary and developmental interactions considering organisms as *active* participants in that process. It is in that sense that organismal individuality is agential. One main focus of Walsh (2015) is why organisms are alienated from evolutionary biology; why they were eclipsed by preoccupation with mechanisms, genes, and population dynamics. Evolution as an ecological phenomenon means that it "happens to a population (or lineage) as a consequence of individual organism's purposive engagement with their conditions existence in the struggle for life" (2015, 208). Of greatest theoretical significance is not the gene or the organism per se, but rather the "organism situated in a system of affordances" (209). In that sense, organisms are *agents* as both subjects and objects of evolutionary change. Walsh (2015), Sultan (2019), and Sultan et al. (2021) are all explicit about their inspirations from Lewontin who said,

> The organism cannot be regarded as simply the passive object of autonomous internal and external forces; it is also the subject of its own evolution (1985, 89).

Thus far, a different conceptual stance was discussed: a Lewontin-inspired call to rethink experimental design according to Sultan et al. That stance designates *active* roles to organisms in evolution through their development (i.e., from an "evo-devo" point of view) as proper subjects, rather than passive objects, which was conceptually eclipsed at some point. Consider two intriguing narratives on how that eclipse occurred.

On the one hand, Walsh tells us that organisms in their ecological contexts (i.e., what they do in pursuit of their ways of life) is a cardinal lesson from Darwin's *Origin* (2015, x). This was obscured by marginalization of organisms under Synthetic theory, which disadvantages evolutionary understanding. That organisms as purposive agents contribute to – *or enact* – evolution is a feature lost to history nearly a century after Darwin (xii). Walsh states, "In Darwin's evolutionary thinking, organisms surrender some, but by no means all, of their theoretical significance" (44). On this reading of Walsh's work, centering agency is a calculated return to Darwin.

---

[53] Sultan et al. identify three gaps without an agency perspective: phenotypic variation, trait transmission from parents to offspring, and the origins of complex novel traits.

On the other hand, historian Jessica Riskin tells us that Darwin wrestled with agency. After Darwin's study of Paley's argument from design at Cambridge, he adopted certain principles, that is, "the notion of mechanical adaptation or 'fitness' of parts, and the related requirement that parts be passive, that a properly scientific account of living phenomena ascribe no agency to the phenomena themselves" (2016, 215). Darwin was deeply torn between the "mechanist dictate to banish agency from nature and the organicist impulse to naturalize agency, to make agency synonymous with life" (2016, 215). This reveals a tension in the history of evolutionary thought, which complicates the return to Darwin. But why?

One explanation is that agency's relationship with purposive action is not unlike "occult qualities" from scholastic science. Scholastic science posited innate tendencies, for example, such as an object's innate tendency to fall once dropped because of the object's drive toward Earth's center. In *Scientific Revolutions,* Kuhn contrasted scholastic science with early Newtonians seeking corpuscle-mechanical explanations of gravity, rather than posit innate qualities. So, accepting agency risks something similar: a scholastic absence of mechanical explanation in favor of occult natures.[54]

In fairness, agency is empirically established by those working within evo-devo: feedback processes in development, and between organisms and their environments, actively contribute to shaping evolutionary change. Sultan et al. (2021, 4) provide an experimental example. When parental anemonefishes are exposed to high concentrations of carbon dioxide in their development, their offspring then develop normally in elevated CO_2, and offspring of unexposed parents demonstrate sharply reduced growth. They summarize agency as "the capacity of a system to participate in its own persistence, maintenance, and function by regulating its structures and activities in response to the conditions it encounters" (2021, 4). While not purely mechanical, agency does not defy scientific explanation.

However, naturalizing agency still attracts some stigma due to what agency historically represents. Consider when Riskin says,

> Darwin rejected both internal (striving) and external (divine) agency as elements of scientific explanations (2018, 215).

In other words, Darwin rejected two sources of agency because internal striving harkened back to the scholastic occult qualities of concern, while external divinity brought in supernatural explanation. In a post-Newtonian era, any

---

[54] Later Kuhn argues that Newtonians eventually accepted gravity as an innate attraction between every pair of particles of matter, occult qualities be damned.

successful view about the cosmos' mechanical nature ought to commit neither offense. Riskin continues about Darwin,

> But he adopted the modes of explanation that each had informed: the genetic or historical mode that went with the notion of an internal (striving) agency, according to which living beings actively transformed themselves over time; and the fitness mode that went with the assumptions of divine agency, according to which living beings were static, passive, designed devices (2018, 215).

Riskin identifies two kinds of inner agency in Darwin's work: inherent tendency of the parts (e.g., *Nisus formativus* or the tendency to multiply) and behavior of the overall organism. She clarifies (2018, 227), "Darwin was abidingly ambivalent of the first sort of agency as he was unwavering with regard to the second," which means Darwin was ambiguous on the subject of innate tendencies, such as the tendency to vary (the causes of variation), as well as inner strivings to "to complexify, to progress, to revert" (233).

And so, one risk of naturalizing agency is rediscovering relics of innate tendencies and design. Modern contexts of agency should remain historically informed due to agency's historical relationships with occult and supernatural qualities.

2. A Tale of Caution: Agency & Biological Individuality Writ Large

Discussion thus far centered on organismal agency in evolution and why it remains controversial. In contrast, our cautionary tale is about the agency – inner striving, direction, and purpose – of biological individuality in evolution *writ large*, which surely seemed innocuous during its inception (but arguably was nocuous later).

For Julian Huxley (1912, 1923), the evolution of agency was significant in his theoretical work. As we'll see, Huxley tied agency with a progress-loaded view: evolution aimed toward "Perfect Individuality" through three different grades. Those grades are discussed later. For Huxley, humans were at the pinnacle of that progress. *Individuals in the Animal Kingdom* (1912) is a bizarre read under the scope of agency in evolution, but downright alarming when Huxley's later role in the history of eugenics is considered. He, like many of his contemporaries, aimed to scientifically manage the evolution of individuality. As discussed later, at times he writes metaphorically about evolution becoming conscious of itself or "evolution's agency" with individuality marking the emergence of an *agential capacity* to control the directions toward which evolution tends. For Huxley, it was a moral imperative for humans as products of that agential capacity to do just that – to scientifically manage and perfect evolution of human individuality.

Next, the scope of that cautionary tale about biological individuality is developed. Huxley was a significant contributor to the Modern Synthesis coining its name. His early view of an agent's active role, specifically human agents, in controlling their own evolution meant that individuality reconstructs the conditions in which populations evolve. Coupled with the evolution of agency – agency as perfecting *individuality* – was an ideology of progress. Was that progress toward perfection theological?

While there is not a single mention of God in the 1912 book, as we'll see, Huxley later wrote a spirited introduction to Pierre Teilhard de Chardin's *Phenomenon of Man* (1958). Huxley declares his independently anticipated claims in Teilhard's book (1958, 12), such as the idea that evolutionary processes should only be described in terms of their direction (rather than origin), the tendency of increasing complexity, and the scientific study and management of human evolution in that context. Huxley maintains that while scientists like himself may find it tough to follow Teilhard's reconciliation of evolution with God, the merging of individual human variety into a unity with the emergent Divinity in "no way detracts from the positive value of [Teilhard's] naturalistic approach" (1958, 19). And so, naturalized agency and Christianity, in the cases discussed next, follow two sides of the same fraught coin: ideologies of "progress" infuse biological individuality, such as the "merging" or ridding of deviant variation that undermines its perfection. Our cautionary tale indeed.

Compound Animals, Zooids, & Individual Perfection

A historical take was just provided regarding ambiguity in Darwin's work: internal (inner striving) and external (divine) agency. Two senses of biological individuality's agency were contrasted: the active (versus passive) natures of organisms and individuality's agency in evolution writ large. The latter is further developed below with Julian Huxley's work as a central touchstone. We'll begin with the sort of view Huxley was arguing against. The following looks back to uncover the conceptual and contextual bases of compounds, life cycles, and "progress" by starting with his grandfather, Thomas Huxley (1825–1895).

1. Biological Compounds: Origins vs. Growth

In the *Origin,* Darwin considered hybrid animals as "two different structures or constitutions having been blended into one" (4th Ed. *Origin,* 1866, 317). However, it wasn't until the 6th edition (1872, 441) that Darwin adds a glossary of scientific terms, which included an explicit definition of "zooid" that aligns with Thomas Huxley's view on individuality. Darwin says,

> [b]y means of eggs and by a process of budding with or without separation
> from the parent of the product of the latter, which is often very different from
> that of the egg. The individuality of the species is represented by the whole of
> the form produced between two sexual reproductions; and these forms, which
> are apparently individual animals, have been called zooids (1872, 441,
> original emphasis).

Thomas Huxley was not happy with Darwin's problem of compound individuality (Elwick, 2007, 128). As Elwick describes, Huxley defined biological individuals not by their independence but "as the entire product of a single sexually fertilized ovum" (2007, 133). Independence as a criterion led to difficult questions: "were the detached and free-swimming sexual parts of marine invertebrates entire individuals, or 'mere organs?'" (128). And so, "Huxley privately began calling each false individual a 'zooid'" and announced in 1851 that the distinction between zooid and individual was founded on a zoological basis and a fact of development" (133). Huxley sought to close off the possibility of plants and animals as compound animals by making the problem of compound individuality a pseudo-issue (Elwick 2007, 149).

To Huxley, the aphid, for example, was a single individual composed of zooids that all budded after an initial act of sexual fertilization (Elwick 2007, 133). The term "zooid" was to disambiguate:

(1) Individuals as components of species starting from "true ova" as single cells undergoing development, rather than
(2) mere reproductive bodies as aggregations of cells.

Huxley's temporal definition concerned the developmental cycle starting with a single cell as one individual. He recognized trouble with his temporal definition – to someone interested in morphology the emphasis on origins was an issue and the morphological and anatomical distinctions between a zooid and individual were not always easy to determine (Elwick 2007, 136). How do we distinguish the production of something new from mere growth of the same? Morphologically the answer is difficult to establish. But Huxley thought that *origin* made a qualitative difference: despite the similarity of shape, for example, such as *Nereis* worms produced by budding versus those emerging from sexually fertilized ova. Distinguishing growth from production of a new individual was the *developmental cycle* as suggested in a later lecture:

> And, in this case, the fact is the Sisyphaean process, in the course of which,
> the living and the growing plant passes from the relative simplicity and latent
> potentiality of the seed to the full epiphany of a highly differentiated type,
> thence to fall back to simplicity and potentiality (Huxley "Evolution and
> Ethics 1893).

This excerpt characterizes Huxley's view about the cyclical containment of individuality where parts share the same fate in the reproductive process only to begin anew indicated by reproductive "bottlenecks" from one cell to many and back again. Huxley used developmental life cycles – what he designated the Sisyphean process – to distinguish growth of the same individual from the origin point of a new one. And to emphasize individuality's incessantly repetitive and recursive nature.

I draw on Elwick's discussion of Thomas Huxley's compound animals, individuality, and zooids as it situates biological individuality historically. Through Elwick's archival work, biological individuality was a factor in the professionalization of the life sciences; a terminological factor carving disciplinary boundaries. Yet, individuality also shaped Huxley's views concerning the importance of *origins*: a technical departure, as Elwick (138) states, from competing views, such as Richard Owen's who argued that morphological similarity was more important than origin.

2. Contextualizing Life Cycles & Compounds

Thomas Huxley's choice to contextualize the repetition of individuality's temporal character as Sisyphaean is striking to philosophical ears: the *Evolution and Ethics* lecture is Huxley's warning against drawing social meaning from biology. Huxley meant to convey the injustice, and even immorality, of nature through a Buddhist framework of meaning (Himmelfarb 2014). However, read through the lens of individuality reveals him grappling with the individual's place in it all: "individual existences are mere temporary associations of phenomena circling round a center, like a dog tied to a post" (Huxley 1893). He muses at the cyclical going forth and returning to the starting point as like ascent and descent, which he viewed as the great leveler: all life forms "from very low forms up to the highest ... the process of life presents the same appearance of cyclical evolution" or the incessant return to origin points (1893, also see footnote 27 of this Element). The beanstalk parable with which the lecture begins indicates that science gets us to high places but has nothing to say about what it all means. One can infer that for Huxley, biological individuality marked a pervasive cycle with no greater social meaning. Biological individuality was the great, repetitive leveler of monotony and evolution's recursive nature was more of a cycle than directional per se: changes leading nowhere in particular as depicted in Figure 8.[55]

However, Thomas Huxley's grandson, Julian, did not share the same outlook.

---

[55] Gould's metaphors of time's cycle and time's arrow are delightfully rediscovered in contrasting the Huxleys' views on individuality over evolutionary time.

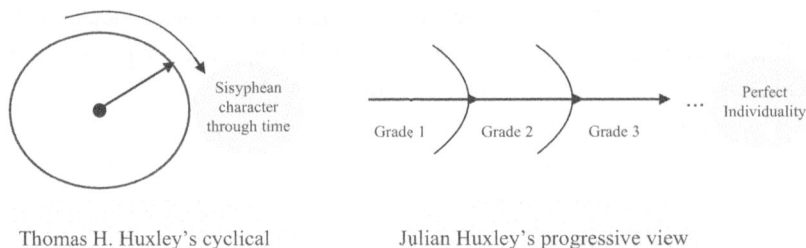

Thomas H. Huxley's cyclical
view of individuality in evolution

Julian Huxley's progressive view
of individuality in evolution

Figure 8 Depiction of Thomas Huxley's cyclical characterization of
individuality through time contrasted against Julian Huxley's view that
individuality progresses along a vector directed toward Perfect Individuality.
While T. Huxley resisted applications of biology to society, J. Huxley's concept
of Perfect Individuality concerns that very application

A young Julian Huxley cited his grandfather's work as an epigraph in the first
essay of *Essays of a Biologist* (1923, 3). Compounds, for Julian were not pseudo-
problems, but rather *emergent* and indicative of directional progress (1923, 242).

Huxley (heretofore Julian – the grandson) argued that new combinations and
properties arise all the time; that compounds, whether chemical or organic, were
not merely reducible to their parts. Citing his own first publication *Individuals
in the Animal Kingdom* (1912), Huxley identifies compound animals, whether
physically bound like Hydra or a Portuguese Man o' war, or "mentally bound"
like eusocial insects, as cases exemplifying intermediate steps of *progress*
(1923, 85). In other words, compound-*ing* is a mode of progress. He states,

> The main fact abides – that progress is an evolutionary reality, and that an
> analysis of the modes of biological progress may often help us in our quest for
> human progress . . . The next great problem on which biology has something
> to say to sociology is that eternal one of the relation between individual and
> community (1923, 90).

There is a complex dimension to how young Huxley identifies individuality
with progress in his 1912 book published a decade or so prior. *Individuals in the
Animal Kingdom* identifies three grades of individuality with compound wholes
as essential building blocks toward the second and third grades (see 1912, 158).
Let's consider all three to unravel Huxley's view of how evolutionary progress
occurs.

3. Individuality as a Driver of Progress in Evolution

The first grade of individuality consists of individuals that, while often homo-
genous, are "marked off" as a closed system (50). Compound individuals

without (e.g., colonial algea like Gonium) and with (e.g., Volvox) division of labor are contrasted against full individuals (e.g., a protozoan or fertilized ovum) of the first grade.

The second grade includes parts of compounds losing their own independence and is "in essence a progress towards greater complexity" through internal differentiation and specialization (136). Compound individuals without (e.g., some sponges) and with (e.g., hydroid colonies) division of labor are contrasted against full individuals in this second grade, namely, humans "regarded singly," and Hydra.

The third grade, in Huxley's appendix, identifies full individuals with examples like ant communities, human society, and lichen. The third – and as such highest – grade involves subordination of lower individualities into higher with symbiotic interdependence as a (sometimes loose) binding factor. Compared to the second level where human organisms are the paradigmatic example, the third level of eusocial colonies and human societies is viewed as immature or rudimentary in its current development overall. But his idea was that *we can manage that progress of human society's evolution by constraining and directing its individual identity.* Huxley identifies the highest common measure of progress as individuality, which establishes "a direction in which its movement is tending, and from that deduce the properties of the Perfect Individual" (1912, 2). This directionality of life is riddled with "limitations of her own physical basis" where imperfect materials only carry so far, but the emergence of the human brain saves the day:

> ... she can go no further forward – the spirit is willing, but the flesh is weak. So far, the range of action has been dependent upon actual mass of substance, diverseness of action upon complexity of substance, and length of action upon duration of substance. Now this direct way is barred: but she finds out another path. She produces a unique type of mechanism ... the human brain ... [a]t once the individuality is released from waiting servile upon substance (1912, 29).

Human consciousness, or agency, transcends substance and thus humanity is meant to further individuality's perfection by manipulating the very processes that gave its existence.[56] The relationship of this view with the social peril of perfection in eugenics will be discussed in more detail later.

Given that Perfect Individuality peaks with humanity's emergence and management of evolution, recall that Huxley's introduction to Pierre Teilhard de

---

[56] Young Julian's reading of meaning and progress into evolution are stamped with his grandfather's quotes. Yet the essence of his views is in striking contrast to Thomas Huxley's warnings about reading value into nature.

Chardin's *The Phenomenon of Man* (1958) was one of praise about their converging ideologies. Citing his own past work concerning the uniqueness of humanity and human evolution and defining evolutionary progress, Huxley (1958, 27) describes himself as independently coming up with two similar ideas: First, that evolution was finally becoming conscious of itself and second, that the evolutionary process led to higher degrees of organization toward becoming One.

For Teilhard, perfection was a complicated concept.[57] In a rather favorable review, Simpson (1960, 207) defined Teilhard's view of perfection as "the consciousness of the universe, which will have evolved through man, [and] will become eternally concentrated at the 'Omega point' . . . [t]he whole process is intended; it is the *purpose* of evolution, planned by the God Who is also the Omega into which consciousness is finally to be concentrated." Citing Teilhard's views on an irreversible perfection (i.e., indicating a vector of progress along a line of time), Simpson notes that book is submitted as a scientific treatise, yet wholly devoted to a thesis that is not demonstrable scientifically. Teilhard's premises were predominantly religious, and the work is not a derivation of religious conclusions from scientific premises (Simpson 1960, 207). In contrast, one might object that Huxley's view of progress toward Perfect Individuality was derived from scientific premises by way of his empirical work. However, how Huxley describes the premises of his own work is more nuanced than that.

In the 1912 preface he states, "I have tried to show in what ways Individuality, *as thus defined by me,* manifests itself in the Animal Kingdom" (viii, original italics). Thereafter, Chapter 1 begins with a quote from Nietzsche's *Zarathustra* concerning the perfection of individuality, to which Huxley proclaims that his own individuality, and individuality in general, is now within the purview of the Zoologist, rather than only the philosopher. But why would the Zoologist take interest in understanding individuality in nature and curating its ideal?

One of Huxley's close colleagues, J.B.S. Haldane, also later characterized the evolutionary process as "passing from the stage of unconsciousness to that of consciousness" in an article outlining the possibilities for radical improvement of humanity's evolutionary future (1947, 51).[58] Haldane identifies biologists as not just mere fact collectors, but as life's *tailors*. Generally, in the post–civil war

---

[57] See McDannell and Lang (1988) for in-depth historical analysis of theological perfection. There is tension between (1) degenerating from a perfect type, for example, in the sense of historically revealed imperfections in modern structures away from the epitome of Creation's design (i.e., falling away from perfection as a ruin of something once whole) and (2) moving toward or approaching God even asymptotically, which was sometimes considered blasphemous. Thanks to historian J. P. Daly on this point.

[58] See Dronamraju (2017) for discussion of Haldane's relationship with the Huxley family.

era, there was intensifying concern of humanity's degeneration after the end of slavery, the arrival of immigrants, and criminality, as well as birth control methods to prevent the "wrong" people from proliferating in human populations. In distinguishing his eugenics (i.e., as the application of biology to society) from that of Hitler's, Haldane argued that when tailors perform measurements and advise adjustments (e.g., for clothes), it is not without disregard for humanity's qualities beyond mere lumps of matter (Haldane 1947, 45). While he recognized the need for diversity to shape the resilience of evolutionary populations, that diversity was of a very particular sort. Haldane (1947, 51) identified birth control and artificial insemination by "great men" as liberal methodologies to reform society's illnesses and systemic poverty. Huxley too wanted to "improve" the human stock achieved through contraception, and artificial insemination by "highly gifted men" (1936, 199). The relationship between Huxley's 1912 book and his later lectures on eugenics in the 1930s and 1960s is discussed later. For now, the driving point is the following: Huxley's stated premise in his 1912 preface concerning individuality's perfection as now within the purview of the Zoologist is not benign.

Huxley resonating with Teilhard's *Phenomenon of Man* is telling from the perspective of John Slattery (2017). Slattery controversially explores Teilhard's connection to eugenics. His archival work reveals Teilhard's views on the inequality of races, the acceptability of violence, and eugenics as a means to perfect humanity. The Omega Point, in Teilhard's own words, depends on the inequality of races and methods to counteract "unprogressive ethnical groups," a view held not just before WWI, but into the last decade of his life after the horrors of concentration camps became known (2017, 75). Teilhard's line of reasoning also included notes on groups of inferior value compared to whites, which he contended were not due to religious belief, but an inferiority *with natural foundation* (74). Teilhard's Omega Point concerned the biological maturing of the human type toward that of more value – it was motivated by the aim to gain control over evolution toward biological purification and perfection (79).

So, while Teilhard saw evolutionary perfection through a religious lens, Huxley's compounding complexity – his grades of individuality – mark the irreversible and progressive path toward Perfect Individuality. Huxley's very clear about his independent arrival to some of Teilhard's conclusions about progress in evolution (see 1958, 12). He reports that it was in *Essays of a Biologist* (1923) where he began defining evolutionary progress. And in those essays, he argues that compound animals were not pseudo-problems, but emergent phenomena signaling progress (1923, 242). There he cites his own 1912 book *Individual in the Animal Kingdom* as identifying grades (of individuality) that exemplify the immediate steps of progress (1923, 85), but that progress was *not* socially neutral:

> The main fact abides – that progress is an evolutionary reality, and that an analysis of the modes of biological progress may often help us in our quest for human progress . . . [and that] [t]he next great problem on which biology has something to say to sociology is that eternal one of the relation between individual and community (1923, 90).

In Huxley's own words then, his prior 1912 work on individuality's perfection and progress was the foundation for the next task of biology: to direct evolution's progress along the vector of perfection through means of social (not just environmental) reform.

In summary, together agency and increasing complexity shape Julian Huxley's grades of individuality as placeholders for directional progress toward perfection. That's in sharp contrast to his grandfather's views concerning life's recursive and repetitive cycling. For Julian Huxley *biological individuality drove progress of a socially charged variety*. The consequences of that view are unpacked next.

What History Tells Us

History tells stories. Biological individuality matters for the narratives it built around progress and humanity's place in the history of life. I submit that individuality was viewed as a marker of progress wrapped in political ideology. And when tapped for eugenics sentiment, the dark side of individuality emerges.

Progress is a structural ideology that some of our best philosophers and scientists have both championed and criticized. Ruse (2019) pulls together contemporary accounts from leading paleontologists and biologists, which advocate for progress in evolution topping humans at the peak of life's history. That is what Ruse's chapter title "Darwinism *as* Religion" refers to: "meaning for the evolutionist is found in the upward rise of the history of life – monad to [human] . . . a teleological force upward to humans" (2019, 111, 122). I maintain that biological individuality had its fair shake in structuring that ideology. Biological individuality's *value* does not only concern the epistemic considerations of theory and methodology previously explored in prior sections. Ideological interpretations of evolution's meaning add controversial (and downright disturbing) social and political dimensions. Exploring the non-epistemic value of biological individuality is not to say that individuality *should* be viewed as a marker of progress, only that it was. Whether it still is, and whether that progress can be *re*conceived as socially neutral, I leave up to the reader to consider on their own.

1. Julian Towing the Line: Individuality as Progress in the History of Life

There is a tension concerning the role of individuality in the history of life. The preceding essay of this Element outlined Julian Huxley's views on individuality, that is, the three grades, in contrast to his grandfather's Sisyphean conception of individuality as representing a continuous, nondirectional cycling. Specifically, the focus was on Julian's notion of Perfect Individuality as one of progress. However, individuality as a progress-marker in the history of life is not an idea that originated with Julian Huxley – he was towing nearly a century-old line. Asa Gray, in his 1861 review of the *Origin,* tells his readers that,

> ... individuality – that very ground of being as distinguished from thing – is not attained in Nature at one leap. If anywhere truly exemplified in plants is only in the lowest and simplest, where the being is a structural unit, a single cell, member less and organ less, though organic,—the same thing as those cells of which all the more complex plants are built up, and with which every plant and (structurally) every animal began its development (Gray 1861, 25).

In other words, "being" as an active quality of organization in nature is individuated by structure, which for Gray was a gradual process often manifest in nested form. He continues,

> In the ascending gradation of the vegetable kingdom individuality is, so to say, striven after, but never attained; in the lower animals it is striven after with greater, though incomplete success; it is realized only in animals of so high a rank that vegetative multiplication or offshoots are out of the question, where all parts are strictly members and nothing else, and all subordinated to a common nervous center,—fully realized, perhaps, only in a conscious person (1861, 25).

According to Gray, *individuality is the goal that nature struggles to achieve only (and finally) reaching its highest pinnacle in humanity.* In other words, for Gray individuality drove ideals of progress in evolution.

As the title of Gray's review suggests, he was arguing that Darwin's theory of natural selection is not inconsistent with theology. That organic life strives for individual perfection is a comment Gray makes while trying to convince his readers that Creator origins do not preclude natural order, and that the record of order implies design. Gray states in reference to biblical creation: "the *pristine* individuals were corporeally constituted like existing individuals, produced through natural agencies" (28, my emphasis).

In other words, to concede that existing individuals were created "after their [pristine] originals" says nothing about what the original types are of, or their mode, or anything else about them. Gray is likely referring to prior views about the independent creation of species; kinds of species were akin to pristine Platonic forms, which are now ruins, mere imperfect caricatures, in their particular instantiations. But what is Gray suggesting about *individuality* as striving for *its* perfection?

Gray's statement, "individuals as created after their kinds" refers to organisms created after their species, but the species are the unchanging and independent categories of nature decreed by God. That is the familiar Great Chain of Being exemplifying an ideological form of progress upward to humans, to angels, to God. Yet Gray is defending Darwin's work that natural connections among species are material and genealogical. *Progress is repurposed*: the patterns of striving for individual perfection are rediscovered through ascending gradations; distinguishing lower from higher according to the degree parts are subordinated to a common purpose of being, as he so defined it. And so, for Gray what would achieving Perfect Individuality be like? By ascending nearer to God. Humanity was created through processes of natural selection in *His* image after all. And since individuality is achieved naturally according to pre-ordained design and marked by the subjugation or dominion over constituents, it marks progress of a very religious (and patriarchal) sort.

Julian Huxley's view of perfection was not far from Gray's. In a cringe-worthy passage from *Individuality in the Animal Kingdom* (1912, 4 my italics) he states, "In this making of Nature his own, civilized man has an individuality vastly fuller, more perfect, *than the savage*. Both in resisting adverse forces and in harnessing the indifferent to his will, he is far superior ... The gradual increase of independence up from the Protozoa to the highest animals ... " is due to increased independence, complexity, and adaptability. That progress for Huxley is, to modern eyes, eugenically packaged, though without God as the driver despite uncanny similarity to Asa Gray's rendition of why individuality matters in the history of life.

2. Eugenics & Individuality: A Tension

Julian Huxley's role as a leading voice of eugenics is complex though: Weindling (2012) identifies him as a crucial bridging figure from the old eugenics (i.e., under state control) to the new eugenics attempting to sustain the scientific management of human evolution as socially progressive, rather than fascist offerings of "solutions" to poverty and disease. For Huxley, individuality's organization was progressively improved through natural selection reaching the pinnacle of a human creature, yet with deficiencies that demonstrated that "unfinished" type (Huxley 1962, 123). He states: "the evolutionary biologist can point out to the social scientist and the politician that this importance of the exceptional individual for psychosocial advance is merely an enhancement of a long-established evolutionary trend" (1962, 129). The 1936 and 1962 Galton Lectures speak for themselves: If *Individuality in the Animal Kingdom* (1912) is the theory of progress through perfection of individuality, those Galton lectures are the how-to guides with concrete suggestions for disincentivizing reproduction of "lower quality" individuals and management of both genetics and the environment as a moral (and even religiously-endowed – see Huxley 1936, 11) imperative.[59]

The foregoing develops the view that individuality was considered an ideological marker of progress in evolution, even if it's not the kind of "progress" that is appropriate: the ideological undertones are dark. Stephen Jay Gould (1989) was concerned with how evolutionary iconography portrayed ideological assumptions exactly like this. Who would have thought that biological individuality had such a significant social role to play? Individuality's role in ideologically charged views of progress takes a different, but no less surprising turn around the Modern Synthesis of the mid-twentieth century. Individuality in life's history becomes aligned with the liberal and democratic values at the time. Consider the following contrast.

George G. Simpson (1902–1984), eminent paleontologist and evolutionist, published a paper titled "The role of the individual in evolution" (1941). He criticized analogies drawn between (1) social groups and individuals and (2) individuals and their organ parts, which both converge on totalitarian ideologies (1941, 15).[60] The view of individuals as primarily subordinate units is viewed as dangerous. He expressed grave concern toward entomologists arguing that social insects do not behave for individual satisfaction. Society as a super-organism or "epi-organism" evolving toward greater integration sets the biologist face-to-face with the totalitarian ideal (16). He chastised reckless biologists elevating the super-organism metaphor to the social realm (18). And for scientifically inaccurate representations of individuals as "pawns of fate" designed toward subservience, rather than their inheritance and environment as of their own making (8). Despite critiquing Fascism and National Socialism as ideologies of "progress" through further subjugation of individuals, Simpson praises Julian Huxley's view as advocating for organisms as more developed than their constituent cells, and that the human society is less individuated than its individual units (19). Individuality in evolution represented progress for Simpson through the furthering of individual satisfaction and breaking free from external control.

And so, in the history of evolutionary thought individuality was viewed as a marker of progress by some serious players – Gray, J. Huxley, and Simpson – who together build a narrative about individuality and evolution's meaning.

---

[59] The racial element shifts from Huxley's 1936 to 1962 Galton Lectures, but the classist interpretation remains of social reform vis-à-vis gifted and exceptional individuals to advance humanity.

[60] See Nyhart and Lidgard (2021) for a full analysis of Simpson's view.

Herein lies a tension, however.

On the one hand, becoming more "individuated," was considered a marker of liberal progress in contrast to fascist and communist ideologies. Becoming more individuated meant increasing autonomy and control over one's environment and the overall direction of their life. That was Simpson's angle.

On the other hand, the aim for individual perfection was tied to the scientific management of human evolution in the form of eugenics practices, which of course manifested as conservative control over bodily autonomy, environmental restrictions, and directing evolution toward a standard of excellence identified only by the elite class under the guise of social reform.

I have no resolution for this tension. Instead, I close with a warning. The historical analyses in this final section are not meant to be whiggish condemnations of past work based on current standards. However, it would be naive and categorically dangerous to assume that social and political motivations did not affect the inner logic of views by prominent figures in the history of biology, even that of Julian Huxley. Taking seriously the role of values in technical scientific reasoning and the inner processes of scientific practice, contemporary presentations of Huxley's scientific work (or of any other figure in the history of biology) must not be read in a value-free vacuum. Individuality has a dark side, but what is to be done about that? This question warrants future interrogation of the tendency to insulate work of biologists from their broader social meanings and motivations. Biological individuality is a case study setting the foundation for exactly that sort of project. The social and political contexts of biological individuality are embedded within the history of eugenics, contexts which reveal how agency, progress, and managing the direction of human evolution together constitute biological individuality's dark side. Lest we forget that dark past: history may not strictly repeat itself, but it can, and sometime does, rhyme.

Closing Remarks

This Element was developed through the lens of both epistemic and non-epistemic values in science. On the one hand, Sections 1 and 2 drew lessons about both the theoretical and methodological aspects of biological individuality's role in the production of scientific knowledge. On the other hand, Section 3 was dedicated to its under-explored social and political consequences. By drawing from historical figures like Darwin, Asa Gray, Thomas and Julian Huxley, and Teilhard, I argued that biological individuality is historically shaped by political and social ideologies about progress and perfection not only with theological overtones, but also in its relation to eugenics as a conceptual tool for controlling humanity's evolutionary future. To that end, a final lesson can be drawn from S. J. Gould.

We cannot observe historical processes at work, in principle. We only see their products (Box 431, Gould Papers). What happens in the past is inferred from both traces left behind (e.g., fossils, artifacts) and what exists now.[61] Building from that, individuals in biology matter because they embody clues of past origins and transformations allowing us to make inferences about past processes. In other words, biological individuals *are the products of history* resulting from causes that brought them about. And as such, they are the things from which we trace patterns of historical processes while at the same time making inferences about what those patterns *mean*. Thus, individuals in biology are both process and pattern-informing as tracking tools, insofar as natural systems exemplify patterns of some order or another. However, their role in facilitating *ideological* patterns of order, that is, politically, and socially charged ideals, while perhaps less obvious at first, is no less intriguing, and I hope to have shown how that can be downright unsettling when set against the history of eugenics. To the philosophers out there: biological individuality is not, and never has been, value-free.

---

[61] Currie (2018) would add more to historical inference than only traces.

Appendix

Table A.1 Individuality puzzles

Puzzle Cases	Individuality Description	Analytics Disrupted, Challenged, or Revised
Holobionts (macro-micro associations e.g., human-gut flora, Squid-*Vibrio* Consortia, aphid-*Buchnera*)	Multiple genomes within physiological "borders," complex parent–offspring relationships (many-to-many), sometimes both spatially and temporally discontinuous	Genetic homogeneity Heritable fitness Physiological unity
Clonal organisms (e.g., quaking aspens, dandelions, *Armillaria)*	Spatially discontinuous, vary in space and time, but similar genetic identity, sometimes physiologically unified (e.g., root nets underground), or spatially separated across miles	Physiological unity Genetic uniqueness Spatial continuity
Eusocial colonies (e.g., ants, termites)	Spatially discontinuous but organizational function as units, phenomenally individuated, but share genetic identity	Physiological unity Genetic uniqueness Functional cohesiveness Boundary closure
Microbial aggregates and colonies (e.g., multispecies biofilms, *Saccharomyces cerevisiae*, *Pseudomona fluorescens*, volvox, lichens)	Functionally organized units, altruistic self-sacrificing cooperation, sometimes connective mediums, for example, shared extracellular matrix	Boundary closure Genetic homogeneity Functional cohesiveness Cooperation & conflict/cheaters
Social amoeba (e.g., Dictyostelium discoideum or "slime molds")	Discontinuity over time and space, generating individuality for a time,	Physiological unity Boundary closure/ Spatial & temporal continuity

Table A.1 (cont.)

Puzzle Cases	Individuality Description	Analytics Disrupted, Challenged, or Revised
	altruistic self-sacrificing cooperation	Cooperation & conflict/cheaters
Colonial organisms/ zooids (e.g., *Physalia* genus, such as Portuguese man o' war, bluebottles)	Multispecies, functionally organized units, division of reproductive labor, sometimes shared mediums, e.g., connective tissues, exoskeleton	Genetic homogeneity Cooperation & conflict/cheaters
Abiotic/biotic associations (e.g., corals: polyps, *zooxanthellae,* and calcite deposits, and Leopold's "land community")	Multispecies crossing levels and abiotic or nonliving components, for example, calcite, soil	Living components Genetic homogeneity

Glossary

Term	Definition
Colonial Logic	Dichotomous or binary perspectives, rather than for example, simultaneously singular and multiple.
Domain-Driven Analyses	Analyses derived from disciplinary domains or subspecialities.
Evolutionary Individuality Category (EI Category)	The category of evolutionary individuality types defined by view of how evolution by selection occurs.
Evolutionary Individuality Concept (EI Concept)	Concepts organizing candidates for evolutionary individuality based on the heredity condition.
Evolutionary Individuality	Units or objects of selection.
Levels of Organization	Levels, nested series of organization ranging from nucleotides, genes, cells, organisms, kinship groups, populations, species, and so on.
Lewontin's Recipe	For populations to evolve by natural selection they must exhibit varying, heritable traits that make a fitness difference.
Major Transitions in Evolution (MTE)	Significant turning points in the history of life.
Ontic	What exists (i.e., objects, concepts, categories, properties, etc.) in a domain (i.e., physical, biological).
Pluralism, Synchronic	A plurality of individuality types at a time.
Pluralism, Diachronic	A plurality of individuality types emerging over time.
Practice-Based Analyses	Analysis of lab and field contexts, for example, methods, community interaction, experimental conditions, and so on.

(cont.)

Term	Definition
Species-As-Individuals Thesis (S-A-I)	The Ghiselin-Hull thesis; species taxa as individuals rather than natural kinds with essences.
Species Category	Defined in contrast to other Linnean classifications (e.g., genus, family, order).
Species Concept	The organization of species taxa according to some set of criteria.
Units of evolution	Objects forming lineages of evolutionary unity ranging over more than one level of organization, usually species
Units of mutation	Objects of mutational processes ranging over different levels of organization, but usually macromolecules.
Units of selection	Objects of natural selection, sometimes referred to as "Darwinian" or "Evolutionary" individuals ranging over different levels of organization.
Value, Epistemic	Knowledge-based values about reasoning, theory, method, success, and so on.
Value, Non-Epistemic	Social, political, moral values that also affect and build knowledge systems.

References

Anderson, J. B., J. N. Bruhn, D. Kasimer et al. (2018). "Clonal Evolution and Genome Stability in a 2500-Year-Old Fungal Individual." *Proceedings Biological Sciences*, 285(1893): 20182233.

Beebe, K., and A. D. Kennedy (2016). "Sharpening Precision Medicine by a Thorough Interrogation of Metabolic Individuality." *Computational and Structural Biotechnology Journal*, 14: 97–105.

Booth, A. (2014). "Symbiosis, Selection, and Individuality." *Biology & Philosophy*, 29: 657–673.

Bouchard, F. (2013). "What is a symbiotic superindividual and how do you measure its fitness?" In eds. F. Bouchard and P. Huneman, *From Groups to Individuals: Evolution and Emerging Individuality*, pp. 243–264. MIT Press.

Bouratt, P. (2015). "How to Read 'Heritability' in the Recipe Approach to Natural Selection." *British Journal for the Philosophy of Science*, 66(4): 883–903.

Box 431 Audio Lecture "Boundaries and Categories." Stephen Jay Gould Papers, M1437. Dept. of Special Collections, Stanford University Libraries, Stanford, CA.

Brandon, R. N. (1999). "The Units of Selection Revisited: The Modules of Selection." *Biology and Philosophy*, 14(2): 167–180.

Bueno, O., R.-L. Chen, and M. B. Fagan (2018). *Individuation, Process, and Scientific Practices*. Oxford University Press.

Buss, L. W. (1987). *The Evolution of Individuality*. Princeton University Press.

Calcott, B., and K. Sterelny (2011). *The Major Transitions in Evolution Revisited*. MIT Press.

Cartieri, F., and A. Potochnik (2014). "Toward Philosophy of Science's Social Engagement." *Erkenn*, 79: 901–916.

Cheung, T. (2006). "From the Organism of a Body to the Body of an Organism: Occurrence and Meaning of the Word 'Organism' From the Seventeenth to the Nineteenth Centuries." *The British Journal for the History of Science*, 39, 142(Pt 3): 319–339.

Child, C. M. (1915). *Individuality in Organisms*. Chicago University Press.

Clarke, E. (2010). "The Problem of Biological Individuality." *Biological Theory*, 5(4): 312–325.

(2013). "Biological Individuality." *The Journal of Philosophy*, CX(8): 413–435.

(2014). "Origins of Evolutionary Transitions." *Journal of Biosciences*, 39(2): 303–317.

(2016). "Levels of Selection in Biofilms: Multispecies Biofilms Are Not Evolutionary Individuals." *Biology & Philosophy*, 31(2): 191–212.

Currie, A. (2017). *Rock, Bone, and Ruin: An Optimist's Guide to the Historical Sciences*. MIT Press.

(2019a). "Mass Extinctions as Major Transitions." *Biology & Philosophy*, 34(2): 1–29.

(2019b). *Scientific Knowledge and The Deep Past: History Matters*. Preprint.

Darlington, C. D. (1958). *Evolution of Genetic Systems*. Basic Books.

Darwin, C. (1866/1872). *On the Origin of Species by Means of Natural Selection, or the Preservation of Favoured Races in the Struggle for Life*. Murray.

Donahue, M. Z. (2017). "How a Color-Blind Artist Became the World's First Cyborg." *National Geographic Magazine* (April 3).

Doolittle, F. (2013). "Microbial Neopleomorphism." *Biology & Philosophy*, 28(2): 351–378.

Doolittle, W. F., and A. Booth (2017). "It's the Song, Not the Singer: An Exploration of Holobiosis and Evolutionary Theory." *Biology and Philosophy*, 32: 5–24.

Doolittle, W. F., and S. A. Inkpen (2018). "Processes and Patterns of Interaction as Units of Selection: An Introduction to ITSNTS Thinking." *Proceedings of the National Academy of Sciences*, 115(16): 4006–4014.

Dupré, J., and M. A. O'Malley (2009). "Varieties of Living Things: Life at the Intersection of Lineage and Metabolism." *Philosophy and Theory in Biology*, I: e003.

Dronamraju, K. (2017). *Popularizing Science: The Life and Work of JBS Haldane*. Oxford University Press.

Elwick, J. (2007). *Styles of Reasoning in the British Life Sciences: Shared Assumptions, 1820–58*. Pickering & Chatto.

Ereshefsky, M. (2000). *The Poverty of the Linnaean Hierarchy: A Philosophical Study of Biological Taxonomy*. Cambridge University Press.

(2022) "Species." In ed. E. N. Zalta, *The Stanford Encyclopedia of Philosophy*. https://plato.stanford.edu/archives/sum2022/entries/species/.

Ereshefsky, M., and M. Pedroso (2013). "Biological Individuality: The Case of Biofilms." *Biology & Philosophy*, 28: 331–349.

(2015). "Rethinking Evolutionary Individuality." *Proceedings of the National Academy of Sciences*, 112(33): 10126–10132.

Fagan, M. B. (2018). "Individuality, Organisms, and Cell Differentiation." In eds. O. Bueno, R.-L. Chen, and M. B. Fagan, *Individuation Across Experimental and Theoretical Sciences*, pp. 114–136. Oxford University Press.

Gerson, E. M. (2017). "Grounded Theory Methodology for the History of Sociology." In eds. S. Moebius and A. Ploder, *Handbuch Geschichte Der*

Deutschsprachigen Soziologie, pp. 259–269. Springer Fachmedien Wiesbaden.

Ghiselin, M. T. (1966). "On Psychologism in the Logic of Taxonomic Controversies." *Systematic Zoology*, 15(3): 207–215.

(1974). "A Radical Solution to the Species Problem." *Systematic Zoology*, 23(4): 536–544.

(1987). "Species Concepts, Individuality, and Objectivity."*Biology and Philosophy*, 2: 127–143.

Gilbert, S. F., and A. I. Tauber (2016). "Rethinking Individuality: The Dialectics of the Holobiont." *Biology & Philosophy*, 31(6): 839–853.

Gilbert, S. F., J. Sapp, and A. I. Tauber (2012). "A Symbiotic View of Life: We Have Never Been Individuals." *The Quarterly Review of Biology*, 87(4): 325–341.

Godfrey-Smith, P. (2009). *Darwinian Populations and Natural Selection*. Oxford University Press.

(2013). "Darwinian Individuals." In eds. F. Bouchard and P. Huneman, *From Groups to Individuals: Evolution and Emerging Individuality*, pp. 17–36. MIT Press.

(2015). "Reproduction, Symbiosis, and the Eukaryotic Cell." *Proceedings of the National Academy of Sciences*, 112(33): 10120–10125.

(2016). "Mind, Matter, and Metabolism." *The Journal of Philosophy*, 113(10): 481–506.

Goldscheider, E. (2009). "Evolution Revolution: Lynn Margulis." *On Wisconsin Magazine* (Fall).

Goodman, N. (1976). *Languages of Art: An Approach to a Theory of Symbols*. Hackett.

Gould, S. J. (1989). *Wonderful Life: The Burgess Shale and the Nature of History*. W. W. Norton.

Gray, A. (1861). *Natural Selection Not Inconsistent with Natural Theology : A Free Examination of Darwin's Treatise on the Origin of Species, and of Its American Reviewers*. Trübner.

Griesemer, J. (1996a). "Periodization and Models in Historical Biology." *New Perspectives on the History of Life: Memoirs of the California Academy of the Sciences*, 20: 19–30.

(1996b). "Some Concepts of Historical Science." *Memorie Della*: 59–69.

(2005). "The Informational Gene and the Substantial Body: On the Generalization of Evolutionary Theory by Abstraction." In eds. M. R. Jones and N. Cartwright, *Idealization XII: Correcting the Model. Idealization and Abstraction in the Sciences (Poznan Studies in the Philosophy of the Sciences and the Humanities*, vol. 86), pp. 59–115. Rodopi.

(2016). "Reproduction in Complex Life Cycles: Toward a Developmental Reaction Norms Perspective." *Philosophy of Science*, 83(5): 803–815.

(2018). "Individuation of Developmental Systems." In eds. O. Bueno, R.-L. Chen, and M. B. Fagan, *Individuation, Process, and Scientific Practices*, pp. 137–164. Oxford University Press.

Griffiths, P. (2006). "Function, Homology, and Character Individuation." *Philosophy of Science*, 73(1): 1–25.

Haber, M. H. (2016). "The Individuality Thesis (3 Ways)." *Biology and Philosophy*, 31: 913–930.

Haldane, J. B. S. (1947). "Evolution: Past and Future." *The Atlantic* (March): 45–51.

Hammerschmidt, K., C. J. Rose, B. Kerr, and P. B. Rainey (2014). "Life Cycles, Fitness Decoupling and the Evolution of Multicellularity." *Nature*, 515(7525): 75–79.

Haraway, D. (1991). "A Cyborg Manifesto: Science, Technology, and Socialist-Feminism in the Late Twentieth Century." In ed. D. Haraway, *Simians, Cyborgs and Women: The Reinvention of Nature*, pp. 149–181. Routledge.

(2016). *Manifestly Haraway*. University of Minnesota Press.

Havstad, J. (2021). "Complexity Begets Crosscutting, Dooms Hierarchy (another paper on natural kinds)." *Synthese*, 198: 7665–7696.

(2020). "Forty Years after *Laboratory Life*." *Philosophy, Theory, and Practice in Biology*, 3(12): 1–34.

Himmelfarb, G. (2014). "Evolution and Ethics, Revisited." *The New Atlantis*, no. 42: 81–87.

Hostiuc, S., M. C. Rusu, I. Negoi et al. (2019). "The Moral Status of Cerebral Organoids." *Regenerative Therapy*, 10: 118–122.

Hull, D. (1965). "The Effect of Essentialism on Taxonomy: Two Thousand Years of Stasis." *British Journal for the Philosophy of Science*, 15: 314–326, 16: 1–18.

(1976). "Are Species Really Individuals?" *Systematic Zoology*, 25(2): 174–191.

(1978). "A Matter of Individuality." *Philosophy of Science*, 45(3): 335–360.

(1980). "Individuality and Selection." *Annual Review of Ecology and Systematics*, 11: 311–332.

(1992). "Individual." In eds. E. F. Keller and E. A. Lloyd, *Keywords In Evolutionary Biology*, pp. 180–187. Cambridge University Press.

Huneman, P. (2014a). "Individuality as a Theoretical Scheme. I. Formal and Material Concepts of Individuality." *Biological Theory*, 9(4): 361–373.

(2014b). "Individuality as a Theoretical Scheme. II. About the Weak Individuality of Organisms and Ecosystems." *Biological Theory*, 9(4): 374–381.

(2017). "Kant's Concept of Organism Revisited: A Framework for a Possible Synthesis Between Developmentalism and Adaptationism." *The Monist*, 100(3): 373–390.

Huxley, J. (1912). *Individual in the Animal Kingdom*. Cambridge University Press.

(1923). *Essays of a Biologist*. Alfred A. Knopf.

(1936). "Eugenics and Society." *The Eugenics Review*, 28(1): 11–31.

(1962). "Eugenics in Evolutionary Perspective." The Galton Lectures. *The Eugenics Review* 54(3): 123–141.

Huxley, T. H. (1893). "Evolution and Ethics: The Romanes Lecture." *Collected Essays IX*, 46–116.

Janzen, D. H. (1977). "What Are Dandelions and Aphids?" *The American Naturalist*, 111(979): 586–589.

Kaiser, M. I. (2018). "Individuating Part-Whole Relations in the Biological World." In eds. O. Bueno, R.-L. Chen, and M. B. Fagan, *Individuation Across Experimental and Theoretical Sciences*, pp. 63–90. Oxford University Press.

Kaiser, M. I., and R. Trappes (2021). "Broadening the Problem Agenda of Biological Individuality: Individual Differences, Uniqueness and Temporality." *Biology & Philosophy*, 36(15): 1–28.

Kendig, C., and B. A. Bartley (2019). "Synthetic Kinds: Kind-Making in Synthetic Biology." In ed. J. R. S. Bursten, *Perspectives on Classification in Synthetic Sciences*, pp. 78–96. Routledge.

Klein, A. (2020). "Philosophy At War: Nationalism and Logical Analysis." *Aeon* (February 3).

Kovaka, K. (2015). "Biological Individuality and Scientific Practice." *Philosophy of Science*, 82(5): 1092–1103.

Krakauer, D., N. Bertschinger, E. Olbrich et al. (2020). "The Information Theory of Individuality." *Theory in Biosciences*, 139: 209–223.

Kranke, N. (2021). "Do Concepts of Individuality Account for Individuation Practices in Studies of Host-Parasite Systems? A Modelling Account of Biological Individuality." *philsci-archive.pitt.edu*. [Preprint].

Lean, C. (2018). "Indexically Structured Ecological Communities." *Philosophy of Science*, 85: 501–522.

Lewontin, R. (1970). "The Units of Selection." *Annual Review of Ecology and Systematics*, 1: 1–18.

Lewontin, R. C. (1985). "The Organism as the Subject and Object of Evolution." In eds. R. Levins and R. Lewontin, *The Dialectical Biologist*, pp. 85–106. Havard University Press.

Libby, E., W. Ratcliff, M. Travisano, and B. Kerr (2014). "Geometry Shapes Evolution of Early Multicellularity." *PLoS Computational Biology*, 10(9): e1003803.

Lidgard, S., and L. K. Nyhart (2017). *Biological Individuality: Integrating Scientific, Philosophical, and Historical Perspectives*. Chicago University Press.

Longino, H. (2002). *The Fate of Knowledge*. Princeton University Press.

Maienschein, J. (2011). "'Organization' as Setting Boundaries of Individual Development." *Biological Theory*, 6(1): 73–79

Margulis, L. (1998). *Symbiotic Planet: A New Look at Evolution*. Basic Books.

Mayr, E. (1970). *Populations, Species, and Evolution: An Abridgment of Animal Species and Evolution*. Harvard University Press.

McConwell, A. K. (2017a). "Contingency and Individuality: A Plurality of Evolutionary Individuality Types." *Philosophy of Science*, 5(84): 1104–1116.

(2017b). *Individuality, the Major Transitions, and the Evolutionary Contingency Thesis*. Dissertation. University of Calgary.

(2020). Old Haunts and New Insights: Review of Individuation, Process, & Scientific Practice. *High-performance Liquid Chromatography*, 42(4): 1–4.

McDannell, C., and B. Lang (1988). *Heaven: A History*. Yale University Press.

Medawar, P. (1957). *Uniqueness of the Individual*. Methuen.

Michod, R. E., and D. Roze (2000). "A Multi-Level Selection Theory of Evolutionary Transitions in Individuality." *Artificial Life VII Workshop Proceedings, Portland, Oregon*: 82–85.

Millstein, R. L. (2018). "Is Aldo Leopold's 'Land Community' an Individual." In eds. R.-L. Chen, O. Bueno, and M. B. Fagan, *Individuation, Process, and Scientific Practices*, pp. 279–302. Oxford University Press.

Molter, D. (2017). "On Mushroom Individuality." *Philosophy of Science*, 84(5): 1117–1127.

Nyhart, L. K., and S. Lidgard (2021). "Revisiting George Gaylord Simpson's 'The Role of the Individual in Evolution' (1941)." *Biological Theory*, 16: 203–212.

Nuño de la Rosa, L. N. (2010). "Becoming Organisms: The Organisation of Development and the Development of Organisation." *History and Philosophy of the Life Sciences*, 32: 289–315.

Okasha, S. (2006). *Evolution and the Levels of Selection*. Oxford University Press.

O'Malley, M. (2014). *Philosophy of Microbiology*. Cambridge University Press.

Osler, W. (1904). *Science and Immortality: The Ingersoll Lecture*. London: Constable.

Pedroso, M. (2017). "Inheritance By Recruitment." *Biology & Philosophy*, 32(1): 127–131.

Pennisi, E. (2018). "'Humongous Fungus' Almost as Big as the Mall of America." *Science Shots: Plants & Animals* (October 10).

Plutynski, A. (2018). *Explaining Cancer: Finding Order in Disorder*. Oxford University Press.

Pradeu, T. (2010). "What is an Organism? An Immunological Answer." *History and Philosophy of the Life Sciences*, 32(2/3): 247–267.

(2012). *The Limits of the Self: Immunology and Biological Identity*. Oxford University Press.

(2016)."Organisms or Biological Individuals? Combining Physiological and Evolutionary Individuality." *Biology & Philosophy* 31(6): 797–817.

Queller, D. C. (2000). "Relatedness and the Fraternal Major Transitions." *Philosophical Transactions of the Royal Society of London. Series B: Biological Sciences*, 355(1403): 1647–1655.

Queller, D. C., and J. E. Strassmann (2012). "Experimental Evolution of Multicellularity Using Microbial Pseudo-Organisms." *Biology Letters*, 9(1): 20120636.

Rainey, P. B., and B. Kerr (2010). "Cheats as First Propagules: A New Hypothesis for the Evolution of Individuality During the Transition from Single Cells to Multicellularity." *Bioessays*, 32(10): 872–880.

Regan, T. (1983). "Animal Rights, Human Wrongs." In eds. H. B. Miller and W. H. Williams, *Ethics and Animals*, pp. 19–43. Humana Press.

Richmond, M. (2001). "The Cell as the Basis for Heredity, Development, and Evolution: Richard Goldschmidt's Program of Physiological Genetics." In eds. M. D. Laubichler and J. Maienschein, *From Embryology to Evo-Devo: A History of Developmental Evolution*, pp. 169–212. Cambridge University Press.

Riskin, J. (2018). *The Restless Clock: A History of the Centuries-Long Argument Over What Makes Living Things Tick*. University of Chicago Press.

Ruse, M. (1987). "Biological Species, Natural Kinds, Individuals, or What?" *The British Journal for the Philosophy of Science*, 38(2): 225–242

(2019). *A Meaning to Life*. Oxford University Press.

Salisbury, D. (2015). "Will the Pronoun I Become Obsolete? A Biological Perspective." *Science Daily* (August 19). www.sciencedaily.com/releases/2015/08/150819120658.htm.

Sampson, T. R., and S. K. Mazmanian (2015). "Control of Brain Development, Function, and Behavior by the Microbiome." *Cell Host Microbe*, 17(5): 565–576.

Santelices, B. (1999). "How Many Kinds of Individual are There?" *Trends in Ecology & Evolution*, 14(4): 152–155.

Sapp, J. (2009). *The New Foundations of Evolution: On The Tree of Life*. Oxford University Press.

Schulenburg, H., J. Kurtz, Y. Moret, and M. T. Siva-Jothy (2009). "Introduction. Ecological Immunology." *Philosophical Transactions of the Royal Society B: Biological Sciences*, 364(1513): 3–14.

Şencan, S. (2019). "A Tale of Two Individuality Accounts and Integrative Pluralism." *Philosophy of Science*, 86(5): 1111–1122.

Simpson, G. G. (1958). "Epilogue: Behavior and Evolution." In eds. A. Roe and G. G. Simpson, *Behavior and Evolution*, pp. 507–536. Yale University Press.

(1960). "Review of *The Phenomenon of Man* by Pierre Teilhard de Chardin." *Scientific American*, 202(4): 201–207.

Sinclair, R. (2020). "Exploding Individuals: Engaging Indigenous Logic and Decolonizing Science." *Hypatia*, 35(1): 58–74.

Slattery, J. (2017). "Dangerous Tendencies of Cosmic Theology: The Untold Legacy of Teilhard de Chardin." *Philosophy & Theology*, 29(1): 69–82.

Smith, J. M., and E. Szathmary (1997). *The Major Transitions in Evolution*. Oxford: Oxford University Press.

Smith, M., J. Bruhn, and J. Anderson (1992). "The Fungus *Armillaria Bulbosa* is among the Largest and Oldest Living Organisms." *Nature*, 356: 428–431.

Sultan, S. E. (2019). "Genotype-Environment Interaction and the Unscripted Reaction Norm." In eds. T. Uller and K. Laland, *Evolutionary Causation: Biological and Philosophical Reflections*, pp. 109–126. MIT Press.

Sultan, S. E., A. P. Moczek, and D. Walsh (2021). "Bridging the Explanatory Gaps: What Can We Learn From a Biological Agency Perspective." *Bioessays*, 44(1): e2100185.

Teilhard de Chardin, P. (1958). *The Phenomenon of Man with Introduction by Sir Julian Huxley*. Harper Perennial.

Van Valen, L. (1976). "Ecological Species, Multispecies, and Oaks." *Taxon*, 25(2–3): 233–239.

Wallace-Wells, B. (2015). "Adventures in the Science of the Superorganism." *NY Magazine* (October 8). www.thecut.com/2015/10/adventures-in-the-science-of-the-superorganism.html.

Walsh, D. M. (2018). "Objectcy and Agency: Towards a Methodological Vitalism." In eds. D. J. Nicholson and J. Dupre, *Everything Flows: Towards a Processual Philosophy of Biology*, pp. 167–185. Oxford University Press.

    (2015). *Organisms, Agency, and Evolution*. Cambridge University Press.

Waters, C. K. (2018). "Ask Not "what is an Individual?" In eds. O. Bueno, M. B. Fagan, and R. Chen, *Individuation, Process, and Scientific Practices*, pp. 91–113. Oxford University Press.

Weindling, P. (2012). "Julian Huxley and the Continuity of Eugenics in Twentieth-Century Britain." *Journal of Modern European History*, 10(4): 480–499.

Wimsatt, W. C. (1972). "Complexity and Organization." *PSA: Proceedings of the Biennial Meeting of the Philosophy of Science Association*, 1972: 67–86.

Zhang, X., and S. Wang (2016). "From the First Human Gene-editing to the Birth of Three-parent Baby." *Science China Life Sciences*, 59(12): 1341–1342.

Acknowledgments

A sincere thanks to Jim Griesemer, Elihu Gerson, Marc Ereshefsky, Joyce Havstad, Adrian Currie, Celso Neto, Makmiller Pedroso, Chris de Teresi, and Luke Breuer for some tough and very helpful discussion and feedback on drafts and ideas, and for the support of UC Davis's Philbio Lab whose members were indispensable during the early writing process including (but not limited to) Roberta Millstein, Natasha Haddal, Denise Hossom, and Tiernan Armstrong-Ingram. I am grateful to my editors Grant Ramsey and Michael Ruse for their patience and advice while undertaking this project, as well as to one referee who provided critical feedback in a positive and constructive way. Another referee provided critiques that ultimately improved the manuscript. Funding for Open Access was provided by UMass Lowell Provost's Office.

For Wade

Cambridge Elements ⸗

Philosophy of Biology

Grant Ramsey

KU Leuven

Grant Ramsey is a BOFZAP research professor at the Institute of Philosophy, KU Leuven, Belgium. His work centers on philosophical problems at the foundation of evolutionary biology. He has been awarded the Popper Prize twice for his work in this area. He also publishes in the philosophy of animal behavior, human nature and the moral emotions. He runs the Ramsey Lab (theramseylab.org), a highly collaborative research group focused on issues in the philosophy of the life sciences.

Michael Ruse

Florida State University

Michael Ruse is the Lucyle T. Werkmeister Professor of Philosophy and the Director of the Program in the History and Philosophy of Science at Florida State University. He is Professor Emeritus at the University of Guelph, in Ontario, Canada. He is a former Guggenheim fellow and Gifford lecturer. He is the author or editor of over sixty books, most recently *Darwinism as Religion: What Literature Tells Us about Evolution*; *On Purpose*; *The Problem of War: Darwinism, Christianity, and their Battle to Understand Human Conflict*; and *A Meaning to Life*.

About the Series

This Cambridge Elements series provides concise and structured introductions to all of the central topics in the philosophy of biology. Contributors to the series are cutting-edge researchers who offer balanced, comprehensive coverage of multiple perspectives, while also developing new ideas and arguments from a unique viewpoint.

Cambridge Elements ⌅

Philosophy of Biology